Cape Cod and the Islands

THE GEOLOGIC STORY

Cape Cod and the Islands

THE GEOLOGIC STORY

Robert N. Oldale

PARNASSUS IMPRINTS

East Orleans, Massachusetts

ISBN 0-940160-53-6

Library of Congress Catalogue Number   91-068255

First Edition

Cover photograph by Dann S. Blackwood,
U.S. Geological Survey

"Here today, gone tomorrow"
Aphra Behn (1640–1689)

To my wife, Gail Carpenter Oldale,
who has been my companion through the years
and who made my days in the field
even more pleasant.

CONTENTS

FIGURES

PREFACE

My first association with the Cape and Islands region was during the middle 1930's when my family camped in a home-made trailer at Popponesset Beach in Mashpee on Cape Cod. Things that I remember are the long ride down a wooded sand road to the camp ground, the smell of the pine trees in the warm sun, swimming, bonfires on the beach at night and answering bonfires on the Vineyard shore, a tour to the top of Cape Cod Light, and a box turtle as a short-term pet.

With an education in hand, I came back to Massachusetts, my native state, in 1955 to work for the United States Geological Survey (USGS). In the fall of 1957, when Sputnik became the first artificial satellite to orbit the Earth, I returned to the Cape, this time with a wife, to begin seismic profiling studies (later the seismic studies were extended to the Islands as well). The Cape had changed some; there were a lot more houses, more paved roads, motels, supermarkets, a two-lane, limited-access superhighway as far as Hyannis, and many more people. Following the formation of the Cape Cod National Sea Shore, I again returned to Cape Cod, with a family, to map the geology. Two colleagues and I mapped from Provincetown to Monomoy Point. Each spring and fall my family and I came to the Cape to continue the mapping from Harwich to Hyannis. Field work during the summer months was impossible because of all the people; it just took too much time to tell them what I was doing. In 1967, because of an increasing interest in offshore geology around the Cape and Islands, I moved my family to Falmouth and joined a cooperative program between the USGS and the Woods Hole Oceanographic Institution. Knowledge gained studying the land geology provided the key to understanding the offshore geology and geologic history based on marine seismic and coring studies.

However, I found that the move to Falmouth was a move from a small rural Massachusetts town to a much larger urbanized town whose population doubled and tripled during the summer. This summer invasion made me think of hibernation, but my wife convinced me that it was the only summer I would get and soon I came to enjoy the summer Cape, tourists and all. Mapping of the Cotuit and Sandwich quadrangles and a geological reconnaissance of the Elizabeth Islands, led to an understanding of the origin of the end moraines, the large glacial ridges that make up the backbones of the Cape and Islands. In 1978, I was given the chance to map the geology of Nantucket Island. Again the mapping was done in the spring and fall, this time accompanied by my wife and two golden retrievers. The famous northeast storm of February 1978 had exposed the geologic section at Sankaty Head, which extended our knowledge of interglacial and glacial history back beyond the last glacial stage.

The outline of the geologic story of Cape Cod and Islands is not new. I have enjoyed adding details to the story. I hope this book will entertain as well as inform the reader and perhaps entice a few young people to study geology and eventually work on the Cape and Islands. We have a great deal yet to learn.

INTRODUCTION

Cape Cod and the offshore islands (Fig. 1), along with Block Island and Long Island to the west, owe their present landscape to the action of the last continental glacier and the rise in sea level that followed. Rock debris was deposited along and beyond the front of the glacier to form the glacial part of the Cape and Islands. The glacial materials have been eroded, transported, and redeposited by the sea to form the Cape and Islands as they are today. The plains, hills, freshwater ponds, and many bays and estuaries owe their origin to the glacier. Other sheltered bays, marshes, cliffs, and beaches owe their origin to the sea. The dunes are shaped by the winds, but their sand comes from a beach or a cliff face and so they also owe their existence to the sea. The role for the continental glacier is over, at least for the time being, but the sea continues to rise and the waves, currents, and winds continue to change the landscape.

The Cape and Islands are to the south and southeast of Boston, Massachusetts (Fig. 1). The Islands consist of the Elizabeth Islands, Nomans Land, Nantucket, and Martha's Vineyard. From Boston, Provincetown is the closest point, about 46 miles, and from Boston, the distance to the Cape Cod Canal, the widest canal in the world, is about 48 miles. Martha's Vineyard and Nantucket lie farther offshore, about 70 miles and 89 miles from Boston, respectively. Cape Cod, in the form of a bent arm, is bordered by Cape Cod Bay to the north and by Buzzards Bay to the west and by Vineyard and Nantucket Sounds to the

1

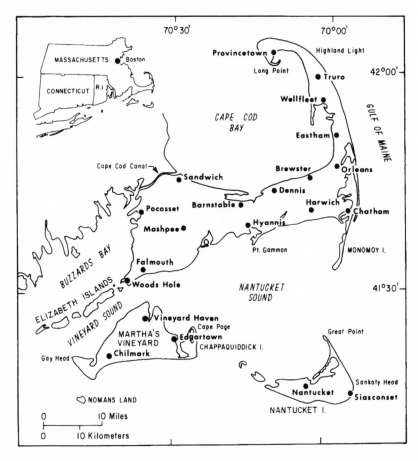

Figure 1.   Index map of Cape Cod and the Islands, with an insert map of southern New England.

south. To the east, the Cape and Nantucket are bordered by the Gulf of Maine, which is a very large embayment of the Atlantic Ocean, and by Great South Channel, a deep-water passage between the Atlantic and the Gulf of Maine that separates the Cape and Islands from Georges Bank. To the south, these islands are bordered by the open Atlantic Ocean.

It is convenient to consider Cape Cod in two parts, the inner cape, or upper arm, known locally as upper Cape Cod, and the outer cape or forearm and wrist, known locally as lower Cape Cod. The terms "upper cape" and "lower cape" hark back to the time of sail, when the prevailing westerly winds caused vessels to sail down wind when going northeasterly and up wind when sailing southwesterly. This sailing terminology is also why a New Englander says "down Maine" even though that state lies to the north.

Cape Cod Bay and Buzzards Bay and Nantucket Sound and Vineyard Sound were shaped by the same geological processes that created the Cape and Islands, and so the glacial and postglacial history of these bodies of water and the Cape and Islands is similar in many ways. Submerged lands such as Stellwagen Bank, north of Cape Cod, and Georges Bank, east of Nantucket, have a geologic history similar to the Cape and Islands, so these features provide us with a preview of the future of the emerged lands if, as expected, sea level continues to rise.

Cape Cod, Nantucket, and Martha's Vineyard are part of the Atlantic Coastal Plain, a physiographic province that lies between uplands and the sea. The Coastal Plain is characterized by a surface of low relief and by altitudes near sea level. It is underlain by a seaward-thickening wedge of unconsolidated sediment that, in turn, is underlain by much older consolidated rocks. Thus, the Cape and Islands more closely resemble the Atlantic coast of New Jersey than the "rock-bound coast of New England" that exists in parts of Connecticut and Rhode Island and north of Boston. The glacial landforms on the Cape and Islands are mostly end moraines and outwash plains (broad flat surfaces formed by glacial meltwater). The landforms built by the sea include barrier spits and islands, which have been deposited by waves and longshore currents and which are capped by dunes (Fig. 2).

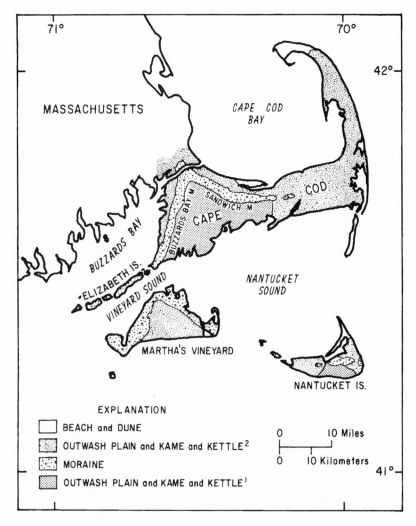

Figure 2. Generalized geologic map of Cape Cod and the Islands showing the locations of the end moraines (M) and adjacent terrains. [1] Outwash plains and kame and kettle terrain older than the adjacent moraine. [2] Outwash plains and kame or kettle terrain younger than the adjacent moraine.

The highest point on Martha's Vineyard is Peaked Hill, at an altitude of 311 feet. On Cape Cod, the highest point is Pine Hill at 306 feet; it is located in the Buzzards Bay moraine in Pocasset. On the Elizabeth Islands, where the Buzzards Bay moraine makes up the entire island chain, the highest altitude occurs on Naushon Island at 174 feet. On Nantucket, the maximum altitude is 111 feet at Sankaty Head. These highest points of land occur on the end moraines, the elongate, lobate ridges marking the glacial terminus at the time they were formed. The outwash plains are lower and have generally flat surfaces that slope gently toward the sea. The barrier beaches and marshes are at or only a few feet above sea level. However, the sand dunes atop the beach deposits generally reach a few tens of feet above sea level, and the dunes on the Provincetown spit reach altitudes of a little more than 100 feet.

The Scope and Organization of This Book

This book describes the geologic history of the Cape and Islands preserved in the rocks and sediments. The story begins with the solid rock, more than 600 million years old, that forms the foundation and ends with the beaches, marshes, and shoals that are forming today. It is written to be accessible to nonscientists and contains a glossary in the back. In the introduction, methods used to study the geology are defined. The body of the book describes and illustrates glacial history and glacial and modern landforms and other features. The book digresses to discuss earthquakes, tides, and storms. Finally the book tries to look to the geologic future of the Cape and Islands. Appendices A–E supplement the text and Appendix F presents a field trip highlighted by overviews of Cape Cod. Appendix G lists other books on the subject.

Topographic Maps

A topographic contour map (Fig. 3) depicts hills, valleys, and other features of the land by using contours to show changes in elevation and slope. A contour is a line of equal elevation that represents the intersection of an imaginary horizontal plane with the ground surface. Contours are much like lake shorelines drawn on a map. A lake surface is a horizontal plane that intersects the lake shore, and its shoreline is a contour line at the elevation of the lake. Contour lines are separated vertically by a constant height, called the contour interval. On the U.S. Geological Survey (USGS) topographic maps of the Cape and Islands (Appendix A), the contour interval is 10 feet, and the values indicate the altitude above mean sea level. Together, the contour lines show the shape of the land, just as the shoreline contour shows the shape of the lake. Spacing of the contour lines shows the slope of the ground. On gentle slopes, they are far apart, and, on steep slopes, they are close together. Closed depressions are distinguished by contours with short right-angle ticks pointing downslope. Underwater contours, also called bathymetric contours, are used to show depths below sea level or below lake level. Because contours show changes in slope of the Earth's surface, they depict the shape of the land or sea floor (Fig. 4).

Figure 3. (opposite)   Topography of the Hyannis 7½-minute quadrangle map (published by the U.S. Geological Survey in 1961), original scale 1:25,000. Topography shown by contours with a 10-foot interval. The map shows the altitude of the land and major natural features such as Sandy Neck, the Great Barnstable Marsh, the Sandwich moraine, Wequaquet Lake, the outwash plain, and Nantucket Sound. It also shows the town boundaries, villages, and roads. The topographic map indicates the location of major cultural features such as the Barnstable Municipal Airport and Hyannis Harbor.

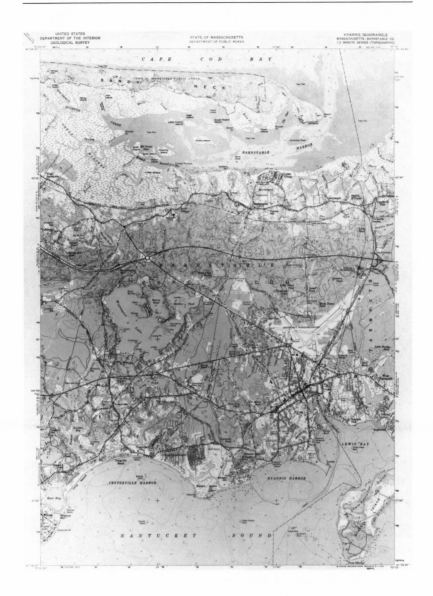

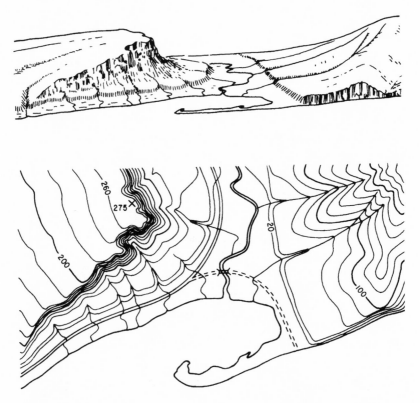

Figure 4. The upper drawing is an oblique view of a landscape composed of a seashore, a river valley, a flood plain, and a barrier spit. To the left of the valley is a steep cliff that borders a flat tableland, and on the right of the valley is a smooth rounded hill. The lower drawing is a contour map, with a 20-foot contour interval, which depicts the same landscape, with the addition of roads and a bridge across the river.

This book includes examples of topographic or bathymetric maps to illustrate the shape or morphology of many geologic features, for example, glacial moraines, outwash plains, beaches, and dunes. In addition to showing the configuration of the land surface, symbols used on topographic maps show other natural features such as lakes and ponds, swamps and marshes, rivers, and bays. They also show the location of roads and trails, buildings, and political boundaries. USGS topographic maps also indicate true and magnetic north and latitude and longitude.

Most of the topographic contour maps published by the U.S. Geological Survey are called quadrangle maps, because they are rectangular and are bounded by lines of latitude and longitude. The size of a quadrangle map is expressed by the number of degrees or minutes of latitude and longitude covered by the map. There are 60 minutes in a degree of latitude or longitude, and the Cape and Islands are covered by twenty-six 7½-minute quadrangles.

Topographic contour maps, as well as other maps such as geologic, hydrologic, and soils maps, are published at various scales (see Appendices A–D). The scale of a map is usually expressed as a ratio, such as 1:100,000, which indicates that one unit of measure on the map represents 100,000 units over the ground. In this case, 1 inch on the map would represent 100,000 inches or about 1.6 miles (Appendix B). Map scales are also shown by bars that represent the length of miles, feet, and kilometers on the map.

Geologic Time

Within the framework of geologic time, the glacial and marine deposits are very young (Fig.5). They overlie much older rock layers. The geologic story of the Cape and Islands begins with the deepest and oldest rock layer and moves upward and forward in time to the youngest of the glacial and marine deposits. The geologic story is very sketchy at first because the rocks are deeply buried and we know them only from geophysical studies and from samples in a few boreholes. Shal-

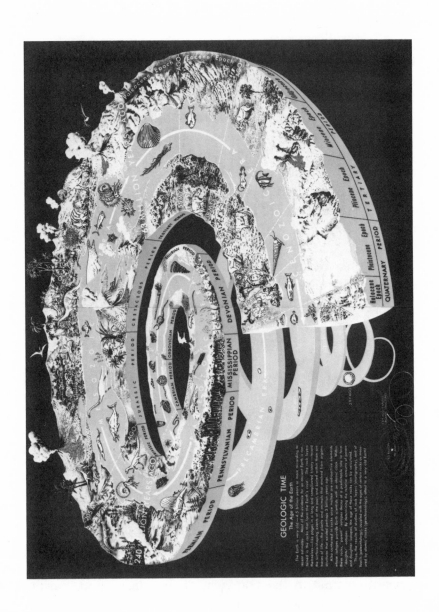

lower rocks are better known because they are locally well exposed in sea cliffs and in gravel pits. Thus, like all accounts of history, the early geologic history of the Cape and Islands is vague, uncertain, and incomplete and becomes more clear, more complete, and more certain as the present is approached. However, even the geologic history over the past 20,000 years is not yet and probably never will be completely understood.

To understand this story, some idea of geologic time is required. The geologic time scale is divided into blocks that range in length from thousands of millions of years to tens of thousands of years. The names for blocks of geologic time are like the names used for historic periods, like Elizabethan or Victorian periods. The geologic blocks of time from the longest to the shortest are called eras, periods, and epochs (Fig. 6) Thus, eras are made up of periods and periods are made up of epochs.

An analogy, based on a similar one previously published by the U.S. Geological Survey, may help us conceive of geologic time. If we were to walk backward in time at the rate of one step per century, the first three and a half steps would return us to about the time of the Pilgrim's landing. To return to the time when the glacier was last in the region, at three feet per step, we would have to walk a little over a tenth of a mile. To return to the time represented by the oldest rocks beneath the Cape and Islands we would have to walk more than 1,000 miles!

Figure 5. (opposite)   Diagram of geologic time from the beginning of the Earth to the Holocene Epoch. On the basis of radiometric dating, Earth is believed to have formed about 4.5 billion years ago. The Precambrian Era represents roughly 90 percent of geologic time whereas the Paleozoic, Mesozoic, and Cenozoic Eras represent about 10 percent. The Quaternary Period represents less than one half of one percent of the history of the earth. Diagram created by John R. Stacy of the U.S. Geological Survey.

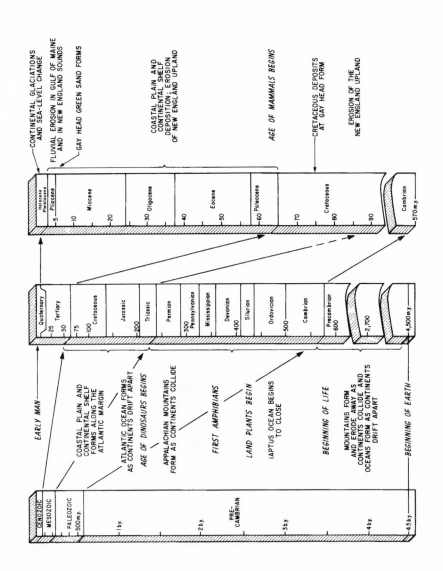

CONTINENTAL GLACIATIONS AND SEA-LEVEL CHANGE
FLUVIAL EROSION IN GULF OF MAINE AND IN NEW ENGLAND SOUNDS
GAY HEAD GREEN SAND FORMS
COASTAL PLAIN AND CONTINENTAL SHELF DEPOSITION; EROSION OF NEW ENGLAND UPLAND
AGE OF MAMMALS BEGINS
CRETACEOUS DEPOSITS AT GAY HEAD FORM
EROSION OF THE NEW ENGLAND UPLAND

Holocene Pleistocene
Pliocene — 5
Miocene — 10
— 20
Oligocene — 30
Eocene — 40
— 50
Paleocene — 60
Cretaceous — 70
— 80
— 90
Cambrian — 570 m.y.

Quaternary
Tertiary — 25
— 50
Cretaceous — 75
— 100
Jurassic
Triassic — 200
Permian
Pennsylvanian — 300
Mississippian
Devonian — 400
Silurian
Ordovician — 500
Cambrian
Precambrian — 600
— 2,700
— 4,500 m.y.

EARLY MAN
COASTAL PLAIN AND CONTINENTAL SHELF FORMS ALONG THE ATLANTIC MARGIN
ATLANTIC OCEAN FORMS AS CONTINENTS DRIFT APART
AGE OF DINOSAURS BEGINS
APPALACHIAN MOUNTAINS FORM AS CONTINENTS COLLIDE
FIRST AMPHIBIANS
LAND PLANTS BEGIN
IAPTUS OCEAN BEGINS TO CLOSE
BEGINNING OF LIFE
MOUNTAINS FORM AND ERODE AWAY AS CONTINENTS COLLIDE AND OCEANS FORM AS CONTINENTS DRIFT APART
BEGINNING OF EARTH

CENOZOIC
MESOZOIC
PALEOZOIC — 500 m.y.
— 1 b.y.
— 2 b.y.
PRE-CAMBRIAN
— 3 b.y.
— 4 b.y.
— 4.5 b.y.

12

Units of geologic time are not of equal length (Fig. 5). The history of the Earth is thought to span 4.5 billion years. The Precambrian, the first era of geologic time, includes about 4 billion years. During this time, life first appeared on the planet. The remaining geologic eras, when life became abundant and highly diversified, represent only about 570 million years of the Earth's history. The Quaternary Period, when man first appeared, is very short when compared with all the other geologic periods (Figs. 5 and 6). The Quaternary Period is still going on and will do so until future geologists can look back and see major changes in planetary geology and life on Earth. By that time it may have lasted as long as some of the other geologic periods, and the future geologists may have evolved to look different from modern man. Future geologists may also see that the Holocene Epoch was a relatively unimportant episode of geologic time, if the continental glaciers once again cover large parts of the Earth. In that case, the Holocene Epoch will be reduced in status to just another episode between two continental glaciations.

Finer time scales are required to unravel the glacial, interglacial, and postglacial history of the Earth (Fig. 7). The time divisions are called glacial stages and interglacial stages. As their names suggest, they are based on classical geologic sections from the Midwestern United States. More recently, a new kind of time scale, based on changes in the content of oxygen isotopes in sea water (an isotope is a variety of a chemical

Figure 6. (opposite)   Geologic time scale. Each succeeding column enlarges a part of the previous column. Ages on the left column are in billions of years (b.y.); on the remaining columns, the ages are in millions of years (m.y.). The times of important events in the development of Cape Cod and the Islands is shown by brackets and arrows.

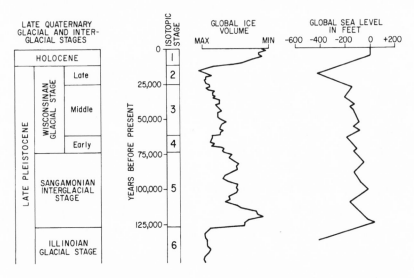

Figure 7.   The glacial (Wisconsinan and Illinoian) and interglacial (Sangamonian) stages of the late Pleistocene and the corresponding marine-oxygen isotope stages. The global ice volume curve is based on the record of the relative abundance of two isotopes of oxygen in sea water. This curve and the curve showing relative sea level clearly show that the last time global ice volumes and relative sea level were close to today's conditions was during the Sangamonian interglacial stage.

element), reversals of the magnetic poles, and radiometric ages (Fig. 7), has provided the detail needed to study the glacial and interglacial history.

Isotope Dating

Seawater contains two oxygen isotopes, oxygen-18 and oxygen-16. The ratio of these two isotopes that were present in the seawater is preserved in the skeletons of microscopic marine organisms that were living at that time. When these microscopic skeletons, obtained from many deep sea cores, are ana-lyzed, it can be seen that the relative abundance of the two oxygen isotopes changed over time. At first, the ratio of these two isotopes was thought to record temperature changes in the oceans, but because glacial ice contains high concentrations of oxygen-16, it was later recognized that the change in the oxy-gen ratio indicated, in large part, changes in global ice volume. The ages of these changes were determined by studying paleo-magnetism (the record of reversals in the magnetic poles of the Earth) and by radiometric dating (the measurement of time based on the rate of decay of radioactive isotopes), and a record of ice volume changes throughout the Pleistocene was ob-tained. Absolute ages, in years before present, for the last glacial stage, the Wisconsinan, the present interglacial stage, the Holocene, and marine oxygen-isotope stages 4 through 1 are largely based on the decay of the isotope carbon-14.

Geologic Mapping

Our understanding of the geology, and consequently the geo-logic history, of the Cape and Islands is based on studies of the strata or rock layers. Some layers are exposed in sea cliffs; others are sampled in boreholes. In geologic mapping, we can determine the relations between geologic units.

Many scientists have studied aspects of the Cape and Islands geology. Among the earliest were Desor and Cabot, who studied the glacial and interglacial deposits exposed in the Sankaty Head cliff on Nantucket (Fig. 1) and published their results in 1849. Since that time, many others have also studied this expo-sure. The most recent study was made by scientists from the USGS, the University of Delaware, and the University of South Florida, following the famous February 1978 northeaster that devastated the Boston area and exposed the beds in the sea cliff.

The Gay Head cliff on Martha's Vineyard (Fig. 1) is another major exposure, which was studied as early as 1823 and most recently during the 1960's and 1970's by Clifford A. Kaye of the USGS. The exposure at Gay Head attracted the attention of many early geologists because it includes strata that are pre-Pleistocene in age and that contain shell, plant, and vertebrate fossils.

A few long boreholes and geophysical studies have revealed the deep subsurface geology of the Cape and Islands. A deep borehole on Nantucket was drilled to a depth of 1,543 feet and ended in solid rock, called bedrock or basement, which provided evidence of the region's role in the development of the early Atlantic Ocean. Seismic soundings were taken on land that determine the speed of sound through various rock layers and their thicknesses, and the depth to basement rock. Magnetic and gravity measurements have provided an additional indication of the nature and structure of the basement rock. Marine seismic soundings were used to identify rock layers beneath the waters adjacent to the Cape and Islands and to provide evidence on the Quaternary history of these nearby offshore areas.

Geologic maps are a primary tool in understanding the geologic history of the Cape and Islands (Appendix C). The first large-scale geologic maps (a large map of a small area) of the Cape and Islands were made by J. B. Woodworth of Harvard University in cooperation with the USGS. The maps used USGS topographic maps at a scale of 1:62,500 (Appendix B) as a base and were published in 1934 by Harvard University as part of an extensive report on the geology of the Cape and Islands (including also Block Island). A second program of large-scale geologic mapping of the Cape and Islands began in 1938, when the USGS made topographic base maps of western Cape Cod at a scale of 1:31,680. A geologic map of Martha's Vineyard by Clifford Kaye was released in 1972 at a scale of 1:31,680, as a USGS Open-File Report. It lacked topographic contours and cultural features. The geologic maps by Woodworth and those of western Cape Cod and Martha's Vineyard are now difficult to obtain.

The USGS resumed geologic quadrangle mapping of Cape Cod in 1964, stimulated by the establishment of the Cape Cod National Sea Shore. The maps used 1:24,000-scale topographic maps as a base for the geology (Fig. 8). These maps cover Cape Cod from Truro to Sandwich and Cotuit and were published as Geologic Quadrangle Maps by the USGS (Appendix C).

A geologic map of Nantucket and nearby islands was made by the U.S. Geological Survey using 1:24,000 scale topographic quadrangle maps as a base. The geologic map was published at half scale (1:48,000).

A geologic map of the Cape and Islands, constructed at a scale of 1:100,000, is a useful complement to this book. The map was compiled from published and unpublished large-scale geologic maps and was printed on a base made from four Landsat images. Unlike a conventional map base, in which the shoreline is standardized at mean high water, this satellite image depicts the shoreline as it was at 10:45 a.m. Eastern Standard Time on August 19, 1978. The geologic quadrangle maps of Cape Cod, the geologic map of Nantucket and nearby islands, and the 1:100,000 scale geologic map of the Cape and Islands can be ordered from the U.S. Geological Survey (Appendix C).

A guide to overviews of Cape Cod from the Cape Cod Canal to Provincetown is included in this book (Appendix F). It will lead the reader to scenic vistas that show how the Cape developed from the time of the last great ice sheet to the present and provide some insight as to the future of the Cape and the islands to the south.

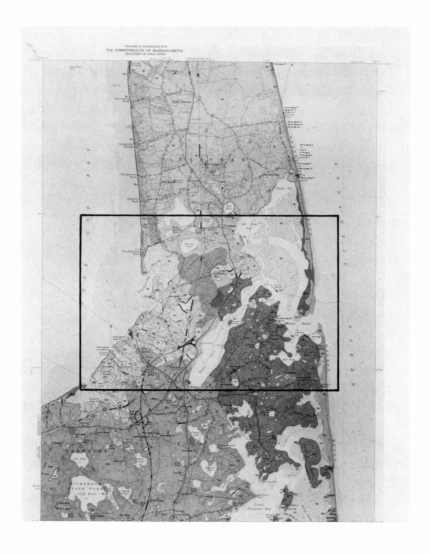

Figure 8. Geologic map of the Orleans quadrangle, original scale 1:24,000, showing the distribution of glacial drift units, the distribution of marine deposits, and dune deposits. The box indicates the area covered by the soils map (Fig. 72). The geologic map was published in 1971 by the U.S. Geological Survey as GQ-931. The authors were R. N. Oldale, Carl Koteff, and J. H. Hartshorn.

THE FOUNDATION

The bedrock lies deeply buried beneath the Cape and Islands (Fig. 9). The bedrock consists of different kinds of consolidated sedimentary, igneous, and metamorphic rocks of Precambrian to Mesozoic age (roughly 600 million to 66 million years old). These rocks are similar to the hard rock found at the surface north and west of Cape Cod. The consolidated sedimentary rocks were laid down layer by layer in ancient streams, lakes, and seas. Some of the igneous rocks (for example, granite) were formed deep in the Earth's crust when plumes of molten rock rose, cooled, and solidified. These deep-seated igneous rocks are called plutonic rocks. Other igneous rocks, called volcanic rock, formed when molten rock reached the Earth's surface, and lava erupted explosively, much as it did during the recent eruption of Mount St. Helens. In other places broad lava flows at the surface cooled to form basalt. Because volcanic rocks and lava flows formed on the surface of the Earth they are called extrusive rocks. During their subsequent geologic history, many of these sedimentary and igneous rocks were deeply buried and deformed by mountain-building forces. Their appearance and mineral composition were changed by heat and pressure to form metamorphic rocks.

We know very little about these oldest rocks beneath the Cape and Islands because they have been sampled only in a few places by drill holes. In Woods Hole, a granite was encountered at a depth of about 270 feet. A deep drill hole in Harwich penetrated schist, a metamorphic rock, at a depth of 435 feet. The schist

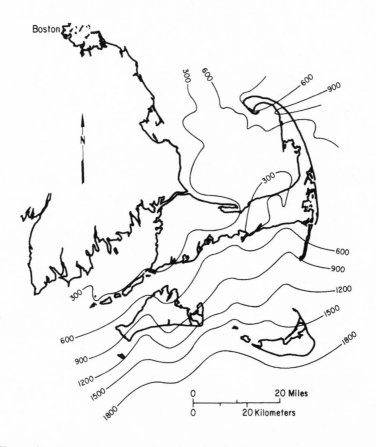

Figure 9. A contour map showing the depth to the bedrock surface beneath Cape Cod and the Islands. The contour interval is 300 feet. Both the lower Cape and the upper Cape are located over highs in the bedrock surface. The bedrock surface beneath Martha's Vineyard and Nantucket slopes gently southward and displays broad valleys that slope in the same direction. These valleys were formed before the Cretaceous and Tertiary coastal plain deposits were laid down.

resembled rocks found in a sedimentary basin of Pennsylvanian age and rocks of Precambrian age in Rhode Island. In Brewster, some distance to the north of the Harwich borehole, a borehole penetrated granite. On Nantucket, a deep borehole (Fig. 10) encountered basalt, a dark-colored igneous rock that cooled

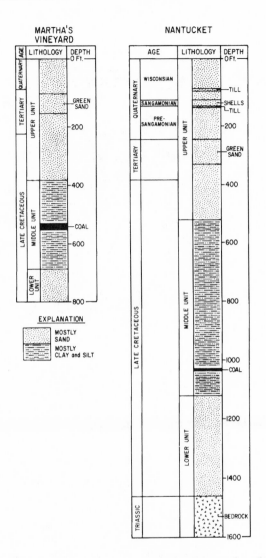

Figure 10. Deep boreholes from Martha's Vineyard and Nantucket. Both boreholes reveal similar geologic sections—an upper sandy unit (shown by the stippled pattern), a clayey middle unit (shown by the dot-dash symbol), which includes a layer of coal called lignite, and a lower sandy unit (also shown by the stippled pattern). The Nantucket boring penetrated basalt of Triassic age, although the Martha's Vineyard boring did not reach bedrock, it was likely very close.

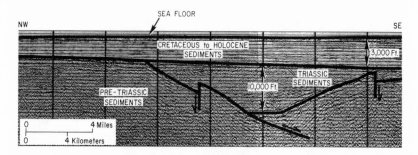

Figure 11.   An offshore seismic profile showing a sedimentary basin containing about 10,000 feet of Triassic strata that were formed as the continents drifted apart and the Atlantic Ocean opened. The seismic line ran to the southeast from just off east central Long Island, NY. Single-barbed arrows show the motion of fault blocks, a result of tension, as the basin formed. The basin rocks are overlain by about 3,000 feet of Cretaceous to Holocene sediment that was deposited along the continental margin, forming the submerged and emerged coastal plain.

from lava that flowed across the land surface from fissures in the crust. The rock resembles basalt from the Connecticut Valley that is found between layers of red sedimentary rock deposited during the Triassic and Jurassic periods.

The basement rocks in the Cape and Islands region have also been investigated using marine seismic profilers, acoustic devices somewhat like a depth sounder, but powerful enough to send waves to penetrate subbottom layers to great depths (Fig. 11). Acoustic profiles show that stratified rocks of probable Triassic to Jurassic age underlie Nantucket (Fig. 12). This interpretation is supported by fragments in the glacial deposits of red sandstone, called arkose, which are similar to arkose of Triassic age in the Connecticut Valley.

Although only a few rock types have been sampled by boreholes, the bedrock beneath the Cape and Islands is probably

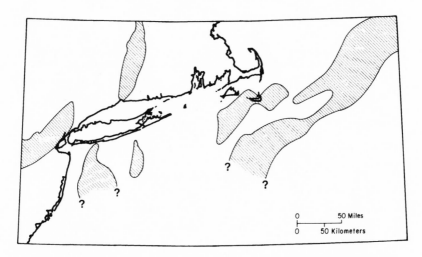

Figure 12. Map showing the onshore and offshore distribution of Triassic and Jurassic basins formed during the initial opening of the Atlantic. The Triassic rocks in the basin south of Cape Cod were penetrated by the Nantucket borehole.

well represented by stones found in the glacial deposits. The most common stones include granite, diorite, gabbro, basalt, and volcanic rocks (igneous rocks). Granite gneiss and quartzite (metamorphic rocks) are also common stones in the glacial deposits. Although some of the stones may have traveled great distances in the ice, most of them probably came from nearby and indicate that the bedrock beneath the Cape and Islands is not significantly different from that elsewhere in southern New England.

The bedrock beneath the Cape and Islands, as in all of New England, represents a very long history of sea-floor spreading and continental drift, as large plates of the Earth's crust drifted around the globe causing oceans to open and continents to collide, closing the oceans and building mountain ranges. Much of the bedrock of this region formed in an early Paleozoic ocean

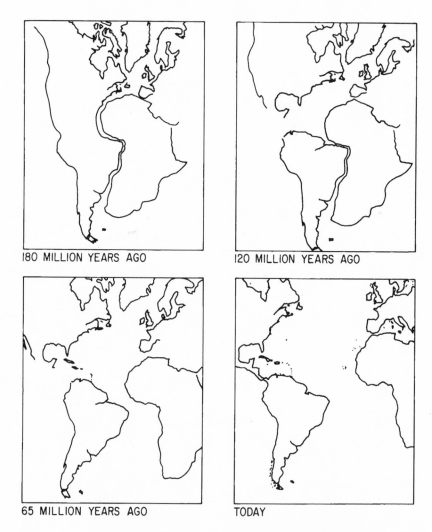

180 MILLION YEARS AGO

120 MILLION YEARS AGO

65 MILLION YEARS AGO

TODAY

Figure 13.   Maps showing the relative position of North America, South America, Africa, and Europe during the opening of the Atlantic Ocean. The stages shown here represent the Jurassic Period (180 m.y. ago), the Early Cretaceous (120 m.y. ago), the Paleocene (65 m.y. ago) and today (based on a drawing by K. O. Emery and Elazar Uchupi).

that existed where the Atlantic does today. This ocean, called Iapetus, closed gradually during the Paleozoic Era, as Africa, Europe, and South America drifted toward North America. As the continents drifted together, compression caused some of the crustal rocks to be thrust downward deep into the Earth, where they melted to form plutonic rocks. Some of the melted rock formed granite deep in the crust. Some was returned to the surface where it was ejected from volcanoes. The colliding continents also caused some of the crustal rock to rise and form mountains. The Appalachian Mountains, including the deeply eroded part found in southern New England, formed as Africa and the eastern part North America collided, closing the Iapetus Ocean.

During early Mesozoic time, the joined continents began to drift apart (Fig. 13), and the Atlantic Ocean had its beginning. The extrusive basalt of early Mesozoic age from the Nantucket borehole and basins filled with sedimentary rock of Triassic age (identified on the basis of geophysical data) provide evidence of the initial opening of the Atlantic Ocean during the Triassic Period. The change from compression as the continents collided to tension as they drifted apart, to form the Atlantic Ocean, brought an end to mountain building in the northeastern United States.

THE COASTAL PLAIN

Over a great period of time, on the order of hundreds of millions of years, erosion lowered the landscape and reduced the relief of the bedrock surface. In southern New England, once-mighty mountains became hills.

Rivers and streams carried sediment eroded from the uplands to the margin of the continent, where it was deposited to form a thick sedimentary wedge that underlies the Coastal Plain, the alluvial surface adjacent to the sea. In the eastern United States, the geographic boundary between the Coastal Plain and the Piedmont is called the Fall Line. Along this line, the rivers crossed from the hard bedrock of the Piedmont to the soft easily eroded sediments of the Coastal Plain. Waterfalls formed along this line and marked the head of navigation of the rivers. Later, during the industrial revolution, the waterfalls became the sites of water-powered mills and eventually of many industrial cities. In New England, the Fall Line pretty well coincides with the coast, and many of the industrial cities are also coastal cities. South of New York City, the Fall Line is well inland, and many of the major industrial cities are far from the sea.

The continental shelf is the part of the Coastal Plain seaward of the shore; however, this distinction is somewhat arbitrary because the shoreline moves back and forth across the continental margin as sea level rises or falls. Thus, the continental shelf is also referred to as the submerged coastal plain as it is in reality. One result of these repeated transgressions and regressions of the sea is an incomplete geologic record because strata laid down during transgression were later, completely or partly, removed by erosion during regressions. Strata that incompletely represent the geologic history of the Coastal Plain on the Cape and Islands are exposed in the cliff at Gay Head on Martha's Vineyard; similar strata were formally exposed on Nonomessett Island, the Elizabeth Island just southwest of Woods Hole (Fig. 1). Before repeated Pleistocene glaciations removed almost all of the coastal plain strata north of the Islands, the inland margin of the coastal plain deposits was quite a bit further north, perhaps near Boston and north of Providence, Rhode Island. At that time, the lay of the land or the geomorphology (the configuration of the Earth's surface) of southernmost New England probably closely resembled the modern Coastal Plain of New Jersey and the Delmarva Peninsula, located between Delaware Bay, Chesapeake Bay, and the Atlantic Ocean.

The coastal plain deposits were cored in the Martha's Vineyard and Nantucket boreholes (Fig. 10). Samples from a deep borehole (Fig. 10) and a water well on Nantucket and from a deep borehole on Martha's Vineyard and the beds exposed by the Gay Head Cliff reveal the nature of the Upper Cretaceous and Tertiary coastal plain strata beneath the Islands.

In a very general way, the sequence of layers in the Nantucket and Martha's Vineyard boreholes can be divided into three parts. The upper part includes Pleistocene and Tertiary strata that are composed mostly of medium to very coarse sand with scattered layers of gravel. The middle part includes strata of Late Cretaceous age that are mostly variegated clay similar to the colorful Cretaceous strata exposed at Gay Head and a thin layer of relatively pure, low-grade coal called lignite. The lower part is also of Late Cretaceous age and is composed mostly of sand containing layers of variegated clay.

The strata in the water well drilled at Coskata were sampled in 1933 by Walter Barrett, then a Nantucket High School student. With the help of J. A. Cushman, an eminent Harvard paleontologist, Barrett studied microfossils from layers of silt and clay in the lower part of the well and assigned a Tertiary age to the strata. His work remained unpublished until 1978, when it was included in a USGS report (Circular 773) on the deep borehole on Nantucket. The deep borehole on Martha's Vineyard did not completely penetrate the lower sandy layer, but bedrock was probably not far below according to seismic work.

The coastal plain strata exposed in the Gay Head Cliff on Martha's Vineyard (Fig. 14) are not in place. They owe their

Figure 14.   Gay Head Cliff, Martha's Vineyard. Most of the strata in this picture are Cretaceous in age and were deposited about 75 million years ago in a setting similar to the present Coastal Plain in New Jersey. The strata are mostly white, gray, red, and black. The beds are not in place, but were thrust forward and upward by the late Wisconsinan ice to form the Martha's Vineyard moraine. A veneer of glacial till occurs at the top of the cliff and is the source of the boulders found on the beach. Vegetated patches on the cliff cap slump blocks that are slowly moving downslope. Photo provided by Peter Fletcher of the U.S. Soil Conservation Service.

position above sea level to the continental glacier that thrust blocks of coastal plain strata southward and upward as it advanced to the island. We know that these beds have been displaced because in the Gay Head Cliff they lie far above their level in the Martha's Vineyard borehole and because they are faulted and folded; in some places, older layers occur above younger layers. However, even though the strata have been pushed from their original position, they provide the best opportunity to study and understand the Late Cretaceous, Tertiary, and early Pleistocene history of the Cape and Islands.

In the Gay Head Cliff, the coastal plain strata are mostly Late Cretaceous in age, the last geologic period in the age of the dinosaurs. The record preserved by these strata represent events that occurred between roughly 80 million and 100 million years ago. Like the Cretaceous deposits in the boreholes, the strata are mostly light colored, but locally, intense colors, including reds, browns, black or dark gray, and green, are present. One thin bed from within the Cretaceous strata has a high content of terrestrial plant fossils including leaves, pine cones, flowers, and seeds. Black layers in the cliff are in some places composed of lignite. Another bed includes mollusk fossils. The plant fossils indicate that most of the strata of Cretaceous age were deposited in streams, lakes, and swamps or marshes on the coastal plain. The fossil marine shells show that at times the coastal plain strata formed in lagoons and embayments along the seashore.

Although the Tertiary strata in the Gay Head Cliff represent only a small part of the overall section, one bed, which can be recognized by its green color (it includes abundant glauconite, a green mineral), contains one of the most interesting fossil assemblages found in New England. Fossils from this bed of greensand were first described in 1793 and were later studied by Sir Charles Lyell, a famous early British geologist. The greensand is Miocene in age and is about 5 million years old. Over many years it has produced the remains of fish, shellfish, crabs, reptiles, and mammals, including whales, rhinoceros, and mastodon.

Perhaps the most intriguing find from the greensand is highly polished pebbles of chert that contain fossil corals of Paleozoic age. The pebbles are thought to be stomach stones, called gastroliths, of seals or walruses. The original geographic source of the pebbles is not known, and bedrock of this type is absent from southern New England. The stones may have been carried by the marine mammals to Martha's Vineyard from the Hudson/Mohawk region near Albany, New York, or from the Hudson Bay region of Canada, where Paleozoic limestones include chert nodules containing fossil corals.

Fossil pollen and spores in the greensand indicate that about 5 million years ago the Cape and Islands region had a subtropical climate and was much closer to the equator than now, as a result of continental drift. Pollen and spores from the youngest Tertiary strata (Pliocene) in the Gay Head Cliff indicate a cool temperate climate that might be a precursor of global cooling and the initiation of glaciation during the Pleistocene.

The fossil content of the strata encountered in the deep boreholes and exposed in the Gay Head Cliff indicate that during much of the Tertiary period the Cape and Islands region was

---

Figure 15. (opposite) Bathymetric map of the Atlantic off southern New England. Bathymetric contours are in meters. One meter roughly equals 3.28 feet. Murray Basin and Wilkinson Basin, in the Gulf of Maine, may be structural and sedimentary basins formed as the Atlantic Ocean opened. Later, the basins were scoured to their present depth as glaciers eroded the soft sedimentary rocks. The Gulf of Maine, Nantucket Sound, Rhode Island Sound, and Long Island Sound represent interior lowlands formed during the late Tertiary by stream erosion and later modified by glaciation. The south shores of the sounds and the north flank of Georges Bank approximately represent the northern limit of continuous coastal plain strata. To the north, the bedrock surface is mostly overlain by glacial drift and Holocene marine deposits. This is part of a 1:1,000,000 scale bathymetric map published in 1965 by the U.S. Geological Survey as I-451. The author was Elazar Uchupi.

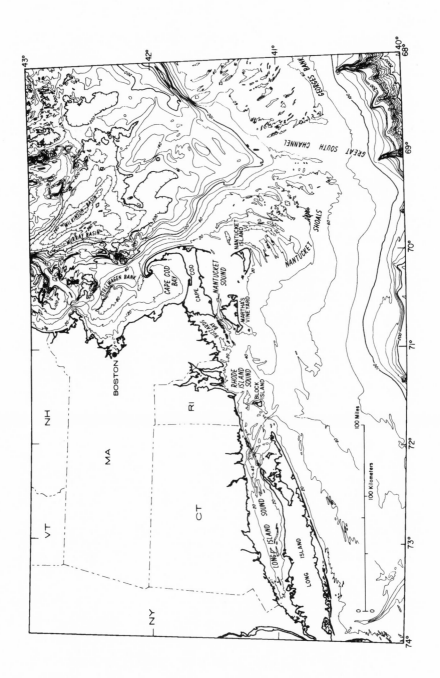

31

submerged beneath the sea and that open continental shelf conditions prevailed. However, from time to time, the sea regressed and the region was exposed to erosion, which resulted in gaps in the Tertiary geologic record. Perhaps the largest gap occurred during the Oligocene Epoch. At that time, sea level was much lower than it is today, and the shore was further seaward. A large east-trending lowland was eroded by streams and rivers; the lowland included all of the sounds adjacent to the south coast of mainland New England and included the Gulf of Maine lowland (Fig. 15). These lowlands were bordered on the seaward side by cuestas, where the gentle seaward slope formed by the accretional surface of the coastal plain was cut by a steep landward-facing erosional slope. Rivers flowed in deep valleys cutting through the cuestas to form water gaps and, thereby, connected the lowlands to the sea. Two such water gaps are Great South Channel east of Nantucket and Northeast Channel at the east end of Georges Bank. Other water gaps through the coastal plain cuesta existed west of the Islands. This landscape of interior lowlands, cuestas, and water gaps was later modified by glacial erosion during repeated Pleistocene glaciations.

OLDER GLACIAL AND INTERGLACIAL STAGES

As the Tertiary Period came to a close, the global climate began to cool, setting the stage for the great ice age, or Pleistocene Epoch, characterized by the growth and expansion of continental ice sheets and mountain glaciers.

The Pleistocene was not a time of perpetual cold global climate, snow, and ice. Instead it was a period when, from time to time, the climate cooled and glaciers began to expand. Ice sheets formed during glacial stages when global climates cooled as the energy from the sun was diminished, mostly the result of cyclical changes in the Earth's orbit and changes in the inclination of the Earth's pole of rotation.

In the high latitudes of the world, more snow falls during the winter than melts during the summer. The snow accumulates layer upon layer, and the deeper layers are compressed to form ice. When sufficiently thick, the lower part of the ice becomes plastic rather than brittle and flows under pressure supplied by the overlying ice and snow. When that happens, the accumulation of snow and ice becomes a glacier.

The times of great glacial expansion are called glacial stages. Alternately, there were periods when global climate was warm, even warmer than today. During these times, the glaciers receded, and, in some places, the continental ice sheets melted away altogether. Times of warming and glacial recession are called interglacial stages.

The present epoch, the Holocene, is most likely an interglacial stage, but only time will tell. Man's impact on global climate, the so-called greenhouse effect, may delay or even prevent the recurrence of a glacial stage. In terms of geologic time, the glacial and interglacial stages are very short. The last glacial stage, called the Wisconsinan in North America, began 75,000 years ago and ended about 10,000 years ago (Fig. 7). The preceding interglacial stage, called the Sangamonian, began about 125,000 years ago.

In southern New England, glaciations interrupted the processes that form the coastal plain. Each of the glaciations probably left a distinctive mark on the landscape, similar to the last one that we can see, a result of the great erosive power of advancing glacial ice and the ability of ice and glacial meltwater to transport and deposit vast amounts of sediment. However, as each glacial advance erases the evidence of the previous advance and retreat, by incorporating older glacial and interglacial deposits into its own deposits, a complete record of the number of glaciations and interglaciations in the Cape and Islands region is not preserved. There is only meager evidence of previous glaciations, and only the last interglacial stage, the Sangamonian, and the last glaciation, the Wisconsinan, have left clear, unmistakable evidence of their happening.

On the Cape and Islands, only Nantucket Island provides convincing evidence of both the last interglacial stage and an earlier glacial stage. At Sankaty Head (Fig. 1), the deposits of the Wisconsinan glaciation lie on top of interglacial marine deposits that are, in turn, on top of glacial deposits. Thus, it is one of the most important Pleistocene geologic sections in New England (Fig. 16).

The marine deposit consists of a basal gravel overlain by an abundantly fossiliferous deposit called the Sankaty Sand. The shells in the Sankaty Sand are living forms and many would be recognized by any shell collector or shellfisherman. Mollusks are the most abundant fossils and include quahogs, steamer clams, mussels, oysters, snails, and whelks. Many bivalves (mollusks with two shells, as in a clam) in the lower part of the Sankaty Sand are in growth position, having both valves together, indicating that they are in the place where they lived and died. Clams and other bivalves in the upper part of the Sankaty Sand occur mostly as single valves and many are broken, indicating that they have been transported and redeposited, perhaps in a beach. Fossil shells and microfossils from the lower part of the Sankaty Sand indicate temperate

Figure 16.   Glacial and marine beds exposed by the Sankaty Head Cliff, Nantucket. The shelly marine beds that occur approximately in the middle of the picture are about 130,000 years old and represent the Sangamonian interglacial stage. Glacial drift below the shelly beds was probably deposited during the Illinoian glacial stage by the next to last glaciation of New England. The drift above the shelly beds is late Wisconsinan in age and was deposited by the Laurentide ice sheet, the last glacier to override southern New England.

seawater, somewhat warmer than that presently off Nantucket, whereas fossils from the upper part of the sand indicate seawater somewhat colder than present. These seawater temperatures probably reflect similar atmospheric temperatures at the time the Sankaty Sand was deposited. A fossil coral from the upper part of the Sankaty Sand was dated radiometrically to be from 128,000 to 140,000 years old. This imprecise age estimate, coupled with the moderate seawater temperatures estimated from the fossils, suggests that the marine deposits at Sankaty Head are Sangamonian in age.

The glacial deposits beneath the marine deposits at Sankaty Head include till (an unsorted mixture of boulder-sized fragments to microscopic-sized clay particles that is deposited directly from glacial ice), glacial outwash (sand and gravel deposited by glacial meltwater), and clay beds that resemble sediments commonly deposited in glacial lakes. The age of the lower glacial deposits is not known; because they are overlain by marine deposits that are probably Sangamonian in age, they cannot be younger than the next to last glacial stage, the Illinoian. The lower glacial deposits are probably remnants of the Illinoian glaciation. The Illinoian glaciation was as great as or somewhat greater than the late Wisconsinan glaciation.

On Martha's Vineyard, the glacial deposits have been assigned to the Nebraskan, Kansan, and Illinoian glacial stages. The oldest glacial deposit underlies a bed containing fossils similar to those found in the Miocene greensand. The fossiliferous bed also contained the bone of an early Pleistocene horse, which indicates that it can be no older than that age. However, none of these deposits have been dated radiometrically, and they may well be out of position because of ice thrusting in the Gay Head Cliff. Such uncertainties make the ages assigned to these deposits questionable.

THE GREAT LAURENTIDE ICE SHEET

Most of the landscape that we see today on the Cape and Islands owes its origin to the last glaciation of New England. The last continental ice sheet in North America formed during the Wisconsinan glacial stage and is called the Laurentide (Fig. 17). Its name is appropriate as the ice sheet that advanced to the Cape and Islands entered New England from the St. Lawrence or the Laurentian region of Canada. The Laurentide ice sheet began to form toward the end of the Sangamonian interglacial stage, less than 100,000 years ago. For much of the Wisconsinan stage, it was confined to Canada or advanced only a short distance into New England. About 25,000 years ago, it advanced across New England and may have reached the Cape and Islands region a few thousand years later. In the more temperate climate of southern New England, the advance slowed and finally stopped in the vicinity of the offshore islands. About 18,000 years ago, as global climate began to warm, retreat began as melting at the glacier margin exceeded the rate of ice advance; retreat continued until the Laurentide ice sheet disappeared altogether (Fig. 18). By 5,000 years ago, the only remnant of the once-great Laurentide ice sheet was on Baffin Island in the Canadian Arctic Archipelago (Fig. 17).

The growth and decay of the Laurentide ice sheet were closely connected to changes in global sea level because the water that

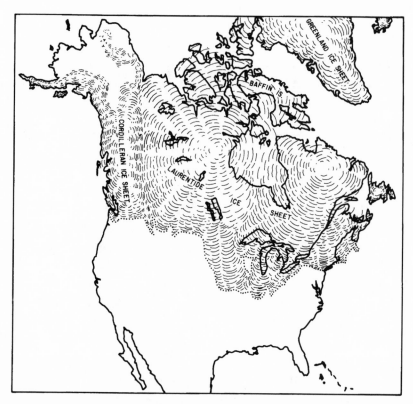

Figure 17. The Laurentide ice sheet in North America during its maximum advance about 21,000 years ago. The Cordilleran ice sheet capped the western mountains and impinged upon Laurentide ice. About 14,000 years ago, the Cordilleran ice sheet separated from the Laurentide to form an ice-free corridor, which would have allowed man to migrate from Asia to North America.

formed the ice was removed from the oceans. Thus, as the Laurentide ice sheet grew, sea level fell, and as it retreated, global sea levels rose. During the maximum growth of the ice sheet, sea level was about 300 feet below present sea level, and the coastal plain south of the Cape and Islands extended to the present edge of the continental shelf, where old shorelines can

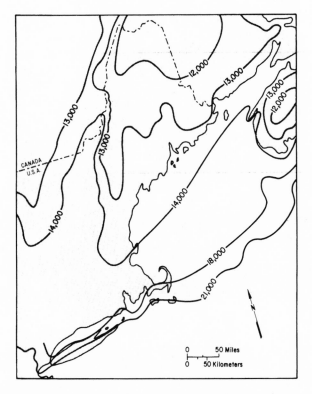

Figure 18.   Stages in the retreat of the Laurentide ice sheet in New England and part of Canada. The lines represent the position of the ice front at the glacial maximum (21,000 years ago) and during the retreat (between 21,000 and 12,000 years ago). Map modified from a 1:5,000,000 scale map published in 1986 by the Geological Survey of Canada and compiled by Arthur S. Dyke and Victor K. Prest.

be seen on the sea floor. When the Laurentide ice sheet had retreated to Baffin Island in northeastern Canada, sea level approached its present level and the lowlands that were to become Vineyard Sound, Nantucket Sound, and Cape Cod Bay drowned. From that time onward, the Cape and Islands began to resemble somewhat the landscape we know today.

At the time of the maximum advance, the Laurentide ice sheet probably had a steep profile as it thickened rapidly northward. When the ice front was at the location of the offshore islands, the ice thickness over Cape Cod was greater than 1,500 feet. In northern New England, the ice was more than a mile thick and completely covered the White Mountains in New Hampshire. Thus, the northward view from in front of the Laurentide ice sheet, as it stood at the Islands, would have been a steeply rising, unbroken expanse of snow and ice as far as the eye could see (Fig. 19, upper). A view to the south would be of broad outwash plains reaching to the horizon (Fig. 19, lower).

A westward aerial view of the ice front, from a vantage point east of the Cape and Islands, would show that the front of the Laurentide ice sheet was not straight, but lobate. The nearest lobe would be located in Great South Channel, the next in Nantucket Sound and Cape Cod Bay, and the third in Vineyard Sound and Buzzards Bay. Appropriately, these lobes are named the South Channel lobe, the Cape Cod Bay lobe, and the Buzzards Bay lobe (Fig. 20). Their placement was influenced by broad shallow depressions in the bedrock surface beneath the ice.

Figure 19. (opposite) Upper photograph: Aerial view along the front of a glacier in Iceland. An up-ice view of the Laurentide glacier, when it stood against the Cape and Islands outwash plains, may have resembled the view in this picture. Lower photograph: A meltwater stream and outwash beyond the ice front of an Icelandic glacier. This view may resemble one that could have been seen beyond the Laurentide ice when it stood at Martha's Vineyard and Nantucket. Photographs by Richard S. Williams of the U.S. Geological Survey.

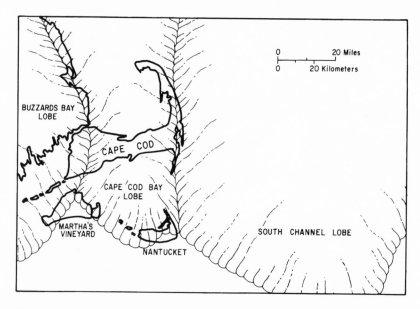

Figure 20.   Lobes of the late Wisconsinan Laurentide ice sheet during its maximum advance in the Cape and Islands region, sometime before 18,000 years ago. Martha's Vineyard was formed by the Buzzards Bay lobe and the Cape Cod Bay lobe (see Fig. 2). Nantucket was formed by the Cape Cod Bay lobe and possibly by the South Channel lobe. Following a retreat of the lobes to the vicinity of Cape Cod, upper Cape Cod was formed by the Buzzards Bay lobe and the Cape Cod Bay lobe. Lower Cape Cod was formed by the South Channel lobe.

GLACIAL SEDIMENTS

The deposits of the Laurentide ice sheet make up the glacial Cape and Islands. The glacial deposits are collectively called drift, a term that was applied when these deposits were ascribed to iceberg rafting during the biblical flood before the theory of continental glaciation. Glacial drift is divided into unstratified and stratified drift. Unstratified drift is laid down directly by ice and generally lacks layering. Stratified drift is deposited by water and is characterized by layers.

As glaciers advance across the landscape, they erode the underlying material, including soil, soft unconsolidated sediments, and hard consolidated rocks; the eroded material is incorporated into the base of the ice. When this unsorted debris, a heterogeneous mixture of boulder-sized to microscopic clay-sized fragments, is deposited by the ice, it is called till (Fig. 21). The debris is carried forward by the advance of the ice and is later deposited in two ways. It may be plastered on the surface beneath the advancing ice or it may be released from stagnant ice by melting at the base or at the ice front. Till also accumulates on the surface of the glacier as the ice near the terminus melts downward. Some of this till may slip from the ice and flow onto accumulating meltwater stream deposits. It is then called flowtill. The alternating processes of erosion and deposition beneath the ice change the stone content of the till. The dominant stones, called clasts, within the till generally reflect the composition of the nearby bedrock. In many places,

Figure 21. Unstratified and unsorted glacial till. Till is composed of all sizes of materials from clay to large boulders deposited directly from the ice with little or no reworking by running water. Entrenching shovel (circled) shows scale; its handle is 21 inches long.

different tills and derivative meltwater deposits can be distinguished on the basis of different stone contents. Typically, till is unsorted (made up of a wide range of grain sizes from clay to boulders) and unlayered, although there are many exceptions, particularly where meltout till is deposited in standing bodies of water.

Melting of the glacier, mostly at the surface and at the base of the ice, provides a vast source of running water. Some meltwater flows across the surface of the ice, but most of it flows downward through the glacier by way of crevasses and tunnels within the ice and then through tunnels at the base of the ice. Water flowing beneath the glacier erodes the till and transports sediment (which consists of all grain sizes from boulders to clay) toward the glacier margin. When the water emerges from the glacier, it forms a melt-water stream. The coarser grain sizes, including boulders, cobbles, pebbles, and coarse sand, are carried along the stream bed by rolling or bouncing. The finer grain sizes (very fine sand to clay) are carried by the water column in suspension. When the sediment carried by the meltwater is deposited against and over the glacier terminus, it is called ice-contact drift. Sediment carried beyond the terminus and deposited downstream is called outwash.

Although the sediment carried by meltwater streams consists of all grain sizes, running water sorts this material. Variations in the speed of water flow sorts the sediments by leaving behind fragments too big to transport. Thus, as the flow slows, meltwater first deposits gravel, then sand, and finally silt and clay. Generally, meltwater streams flow fastest as they emerge from the glacier and slow gradually with increasing distance from the ice margin. As a consequence, meltwater stream deposits are coarsest near the ice, where they include large boulders and coarse gravel (Fig. 22), and finest away from the ice, where they are mostly fine gravel and sand (Fig. 23).

Changes in volume and velocity of meltwater streams occur seasonally. The streams have little volume and flow slowly during spring, fall, and winter when the weather is cold and melting is slow or stops completely. During the summer, when melting is at the maximum, meltwater streams run full and fast. Changes in meltwater volume and stream velocity also occur daily during the melting season. Glacial streams run fast during the afternoon and early evening as the temperature rises and melting increases; they run relatively slowly during the night and early morning as the temperature falls and melting slows. Most of the time the meltwater environment on the Cape and Islands was undoubtedly much like this general picture. However, the seasonal and daily changes in meltwater

Figure 22.   Ice-contact drift of the Malaspina Glacier, Alaska. Poorly sorted stratified drift was deposited against and over the glacial terminus or around ice blocks. In this photograph, the vertical cliffs are made up of glacial ice. Ice-contact drift is composed of stratified drift and in places flowtill, a till that traveled as a mudflow off the surface of the ice and onto the stratified drift. Photograph is from the Malaspina Glacier, Alaska, and was taken by Joseph H. Hartshorn of the U.S. Geological Survey.

Figure 23. Stratified glacial drift exposed in a gravel pit in the Eastham outwash plain. These sediments were deposited by braided meltwater streams and are characterized by lenses of crossbedded sand and gravel left behind as the braided streams changed course. Entrenching shovel (circled) shows scale; its handle is 21 inches long.

discharge were probably interrupted many times by catastrophic releases of great volumes of water from beneath the ice. Such great floods were caused by the sudden breakout of meltwater trapped beneath the ice.

The seasonal and daily changes in stream flow cause most meltwater streams to be braided. As meltwater stream flow slows, the stream drops much of its sediment load to the stream bed. Later, when meltwater stream flow increases, the stream finds a new course around the previously deposited material. As a result, meltwater streams are characterized by numerous shallow broad channels separated by gravel bars that form a braided pattern (Fig. 24).

In vertical sections such as gravel pit walls or sea cliffs, braided stream deposits are typically made up of beds and lenses that are cross stratified, that is the layers of sand and gravel within the bed or lens slope relatively steeply in a down-current

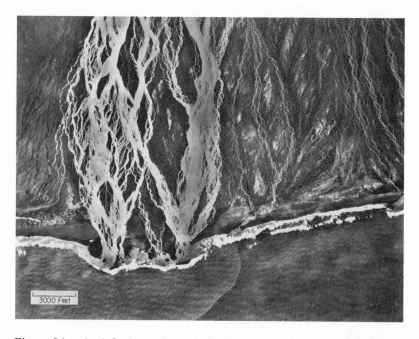

3000 Feet

Figure 24.   Aerial view of a braided outwash plain in southern Iceland. The braided pattern is characteristic of meltwater streams because of daily and seasonal variations in the volume of meltwater and variations in the ability of the streams to transport sediment. These variations are caused by changes in the rate of glacial melting and thus the volume of meltwater. The meltwater streams are entering the sea, where an ephemeral delta is being formed. The lower Cape outwash plains may have looked very much like this as meltwater streams built the outwash plain deltas into Glacial Lake Cape Cod. Photograph supplied by Richard S. Williams of the U.S. Geological Survey.

direction. In many places, the cross stratification occurs in overlapping lenses of sand and gravel (Fig. 23). These lenses are the result of depressions cut in the stream bed by water flow that are quickly filled by sediment.

Where a meltwater stream enters a standing body of water, either a lake or the sea, it slows quickly and deposits its remaining load of coarse sediment at the shore to form a delta (Fig. 25).

Figure 25.   This outwash delta in coastal Maine was formed when a meltwater stream entered the sea. The delta's marine origin is indicated by fossiliferous glacial marine deposits that interfinger with the delta deposits. The sloping layers are called foreset beds and were formed in the sea along the steep face at the front of the delta. The nearly horizontal beds at the top were deposited by a glacial stream carrying sand and gravel to the delta front. The horizontal layers are called topset beds. The outwash deltas on Cape Cod have a similar internal structure, which provides evidence that a glacial lake existed in Cape Cod Bay. Photo by Joseph T. Kelley of the Maine Geological Survey.

Fine sediment carried in suspension enters the standing water and slowly settles to the bottom to form layers of fine sand, silt, and clay. Glacial lake deposits commonly consist of repeated layers of very fine sand and silt capped by clay. Widely scattered boulders to pebbles within the lake deposits are called drop stones and have come from melting icebergs in the lake. If each bed, a couplet of a silty layer capped by a clay layer, in the lake deposit represents one year in the history of the glacial lake, the bed is called a varve (Fig. 26). The lower layer of the varve couplet consists of fine sand and silt that fell to the

Figure 26.   Varves formed in a glacial lake. Each varve couplet consists of two layers. The thicker layer at the bottom of the varve represents summer deposition and consists of mostly silt with some sand and clay. The thin upper layer of the couplet represents winter deposition and consists of mostly clay. Entrenching shovel shows scale; blade is 8½ inches long. Photograph by Gail Ashley, Rutgers University.

bottom during the summer leaving the clay-sized particles in suspension. Later, during the winter, the clay particles settle slowly to the bottom to form the upper layer of the couplet. By counting these annual beds, the life span of the lake can be determined. Varves from a number of lakes can be compared and matched much like tree rings. Using varves from a series of lakes that formed sequentially, the rate of glacial retreat can be determined. If the age of some of the varves can be determined by radiocarbon dating, an absolute chronology of ice retreat can be obtained.

Many meltwater streams on Cape Cod entered glacial lakes that occupied Nantucket Sound and Cape Cod Bay. Some meltwater streams on the islands may also have entered lakes, but most of the streams probably flowed across the coastal plain to enter eventually the sea near the outer edge of the continental shelf. Thus, much of the outwash on Cape Cod is deltaic in origin. Deltaic sediments are mostly sand and contain minor amounts of fine gravel. The beds within the delta slope steeply, as much as 30°. These steeply dipping beds, called foresets, are those that were deposited on the delta front as it advanced into the lake. The foreset beds are capped by gently dipping, almost flat beds called topset beds, generally composed of coarser sand and gravel, that were laid down by the stream as it ran across the foreset deposits to enter the lake along the delta front. The contact between foreset and topset beds marks approximately the level of the lake.

Fine sand and silt on the outwash plains and on the floor of drained glacial lakes were exposed to strong winds blowing off the glacier. These winds picked up the sand, silt, and clay and deposited them atop till and stratified drift surfaces. This uppermost deposit, called the eolian layer, is up to a few feet thick; it is widespread on Cape Cod and the Islands. Although it is not strictly a glacial deposit, it is derived from glacial

Figure 27.   Wind-polished stone or ventifact. These fluted, faceted, and pitted stones were sandblasted by wind-driven sand and silt grains as they lay on the outwash plain surface. Later, frost action worked the ventifacts upward into the eolian deposit that covered the outwash. The unusual shapes of some ventifacts cause them to be mistaken for Indian artifacts. Photograph by Dann S. Blackwood of the U.S. Geological Survey.

debris. As the wind-driven grains sandblasted stones at the surface of the drift, the stones became fluted and polished to form ventifacts (Fig. 27). Frost action has moved the ventifacts upward into the eolian layer. Many ventifacts have unusual shapes, and some have been mistaken for Indian artifacts.

GLACIAL LANDFORMS

First-order glacial landforms, although modified by marine erosion and deposition, include the triangular shape of Nantucket and Martha's Vineyard, the bent-arm shape of Cape Cod, and the arc of the Elizabeth Islands. These landforms owe their existence to the lobate margin of the Laurentide ice sheet (Fig. 20) during its maximum advance and early retreat. Nantucket was formed in the interlobate angle between the South Channel and Cape Cod Bay lobes, while Martha's Vineyard formed in the interlobate angle between the Cape Cod Bay lobe and the Buzzards Bay lobe. Western Cape Cod also formed between these two lobes, while the forearm of outer Cape Cod was deposited along the north-trending front of the western side of the South Channel lobe (Fig. 20). The arc of the Elizabeth Islands mimics the eastern side of the Buzzards Bay lobe as it paused to build the Buzzards Bay moraine during its retreat northward (Fig. 2).

The large bays and sounds also owe their primary shape to the lobation of the Laurentide ice sheet. Nantucket Sound and Cape Cod Bay occupy the depression that once contained the Cape Cod Bay lobe, and Vineyard Sound and Buzzards Bay occupy the depression that contained the eastern part of the Buzzards Bay lobe.

Various glacial processes and depositional environments produce distinctive landforms that provide information on the composition, structure, and relative age of the underlying deposits. The most common glacial landforms on the Cape and Islands are end moraines, outwash plains, and kame and kettle terrain.

End Moraines

End moraines form at the ice front and parallel the ice margin. Prominent end moraines occur along the northwest side of Martha's Vineyard (Martha's Vineyard moraine), in the eastern part of Nantucket (Nantucket moraine), along the shore of Buzzards Bay, including the Elizabeth Islands (Buzzards Bay moraine), and on inner Cape Cod parallel to the Cape Cod Bay shore (Sandwich moraine) (Fig. 2).

The end moraines of the Cape and Islands are segments of two long east-trending moraine belts. The southern belt loops across the islands from western Long Island, New York, to Nantucket. The northern belt loops across Long Island, eastern Connecticut, southern Rhode Island, and inner Cape Cod (Fig. 28). The southern moraine belt generally marks the maximum advance of the Laurentide ice, and the northern belt marks readvances of the Laurentide ice during its overall retreat from New England.

End moraines are broad, elongate, and hummocky ridges that generally stand well above adjacent glacial surfaces. They generally have a steep distal slope (the side of the moraine away from the ice front) and a somewhat gentler slope on the proximal side (the side nearer the ice front). End moraines have a very complex topography made up of ranks of smaller ridges, elongate lows between the ridges, both aligned with the trend of the moraine, and numerous closed depressions of various sizes (Fig. 29). Innumerable boulders, some very large, lie scattered about the moraine surface.

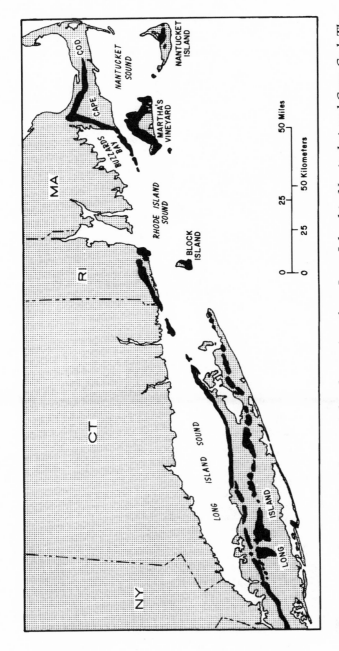

Figure 28. Map showing the coastal end moraines from Long Island to Nantucket and Cape Cod. The southern moraine belt was formed during the maximum advance of the Laurentide ice sheet. The northern belt was formed later by minor readvances of the ice front during the early stage of overall glacial retreat.

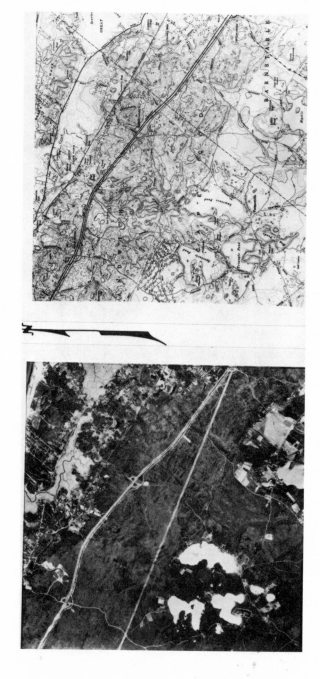

Figure 29. Aerial photograph and part of the Sandwich 7½-minute topographic quadrangle (original scale 1:25,000) covering the same area and showing the grain in the Sandwich moraine. The grain is formed by small ridges formed as push moraines or dump moraines superimposed on the larger thrust ridges that make up the main moraine. The ridges may represent small fluctuations of the ice front during the formation of the Sandwich moraine.

In many places, such as the Midwestern United States, moraines are composed mostly of till, deposited in conveyor belt fashion as unsorted debris carried by the ice is dumped along the glacier margin as the ice melts. However, the great end moraines of Cape Cod and the Islands and to the west on Block Island and Long Island were not formed in this manner. They were formed when the glacier advanced across a bed of unconsolidated sediment, either coastal plain strata, older glacial deposits, or its own glacial deposits. These moraines are formed of sheets of unconsolidated material that were displaced by the weight of the overriding ice and by the advance of the ice. As these sheets were folded and thrust faulted they were forced forward and upward beyond the ice front (Fig. 30) to form the lesser ridges, interridge lows, and some of the closed depressions characteristic of the moraine surface (other closed depressions, called kettles, were formed by the melting of enclosed ice blocks). Thus, the large coastal end moraines were formed glaciotectonically by processes more like those that build mountains than the more passive processes of ice advance and ice front melting (Fig. 31).

The evidence for glacial dislocation of beds is clear in the Gay Head and Sankaty Head Cliffs. In the Gay Head Cliff, the Martha's Vineyard moraine is made up of dislocated and deformed sheets of Cretaceous and Tertiary strata that are stacked one atop the other, in some places with older strata above younger. In the Sankaty Head Cliff, the Sankaty Sand within the Nantucket moraine is far above its proper location, as indicated by the Nantucket borehole, where a shelly bed correlated with the Sankaty Sand occurs at a depth of 112 feet below sea level. In one place in the Sankaty Head Cliff, the Sankaty Sand occurs twice because it has been thrust forward and over itself (Fig. 32). On Nonomessett Island of the Elizabeth Island chain, dislocated Cretaceous and Tertiary strata in the Buzzards Bay moraine were once exposed in a sea cliff. However, this exposure has

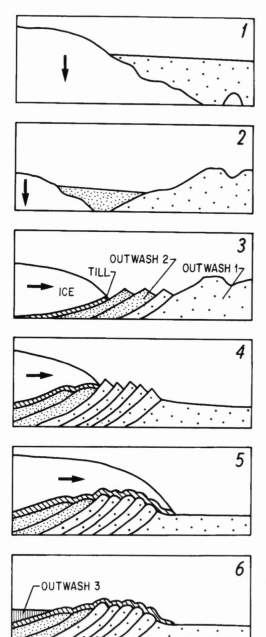

Figure 30.   A model for the formation of the coastal end moraines. Steps 1 and 2 represent the formation of outwash plains beyond a downwasting stagnant ice front, shown by the vertical arrows. Steps 3 and 4 represent the formation of the thrust moraine by the advancing ice front, shown by the horizontal arrows. The thrust moraine is formed by adding thrust sheets at the base of the moraine. Step 5 shows the advancing ice front overriding the moraine and depositing a thin veneer of till atop the moraine. The final step represents the end moraine following ice retreat.

Figure 31. These two cartoons illustrate the formation of the glaciotectonic end moraines in contrast to conventional ideas of moraine formation. In the upper cartoon, the glacier is vigorously advancing over older deposits, piling up thrust sheets beyond the ice front represented by the bulldozer. In the lower cartoon, the conveyor belt represents an ice front that is stationary as ice advance is balanced by melting along the ice front. Debris carried forward by the ice is dumped at the ice front as it melts. Although parts of the coastal end moraines may have been formed by the latter process, most of the moraines were formed by ice thrusting. Cartoons by Carol Parmenter of the U.S. Geological Survey.

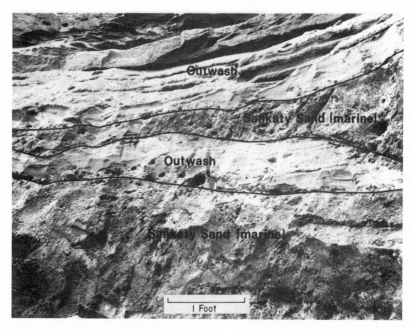

Figure 32.   Evidence of glaciotectonic thrusting in the Nantucket moraine at Sankaty Head Cliff. The shelly marine deposit of the Sankaty Sand of Sangamonian age occurs twice, separated and overlain by outwash of late Wisconsinan age. The upper sheet of Sankaty Sand was thrust over the outwash as the moraine was formed.

been eroded and the only evidence remaining are fragments of fossiliferous sandstone in the cobble beach. Elsewhere, the Buzzards Bay and Sandwich moraines are composed of glacial deposits that are mostly stratified. Evidence that these deposits are not in their original positions is that the lake beds and possible delta beds are found far above any possible lake basin. Thus, they were deposited elsewhere and were later overridden and dislocated by a readvance of the Laurentide ice. Stratified drift and coastal plain deposits within the coastal end moraines are, in most places, capped by a veneer of bouldery till a few feet to a few tens of feet thick (Fig. 33) that was laid down when the glacier overrode the moraine (Fig. 30).

Figure 33. Till over stratified drift in the Sandwich moraine. The till veneer was deposited by the Cape Cod Bay lobe when it overrode thrust sheets that make up the bulk of the moraine. Bulldozer cab (circled) shows scale.

Outwash Plains

Outwash plains are the most common glacial landform on the Cape and Islands. Two large outwash plains occur on Nantucket. A single outwash plain makes up much of Martha's Vineyard. Upper Cape Cod is, in part, made up of three outwash plains and the lower Cape consists mostly of three outwash plains.

Outwash plains are broad, flat, alluvial surfaces, formed by braided meltwater streams, that slope gently away from the position of the former glacier. They are underlain by stratified drift, mostly gravelly sand. The meltwater sediments in the upstream ends of many outwash plains were deposited against and over the front of the glacier. When the glacier melted back, the deposits and the outwash plain surface collapsed to form what is known as an ice-contact head, a surface of irregular topography that, in most places, slopes steeply toward the former position of the ice margin. Thus, the ice-contact slopes mark former positions of the ice margin in much the way that end moraines do.

Deposits beneath the ice-contact slope reflect the close proximity of the glacier and are highly variable in composition. They include stratified glacial drift that ranges from coarse sand and gravel to silt and clay. Locally they include till deposited directly from the ice or as flowtills. Small to very large boulders are common within the ice-contact deposits and are scattered about on the ice-contact slope. Some meltwater streams emerged from the ice and passed through end moraines before they dumped their sediment load to construct a broad outwash plain. These outwash plains lack a well-defined ice-contact head.

The surfaces of the outwash plains are, in many places, interrupted by closed depressions (kettles) that mark the sites of ice blocks buried by the outwash deposits. Some outwash plains include many kettles and are termed pitted (like the Mashpee pitted plain). The ice blocks may have been completely buried and contributed little material to the outwash. Others were only partly buried, and, as they melted, they contributed boulders and minor amounts of sediment to the outwash plain deposits. The eastern outwash plain on Nantucket and the Martha's Vineyard outwash plain formed beyond the maximum advance of the Laurentide ice and therefore lack large kettles.

In places, the outwash plain surface is cut by valleys now unoccupied by streams and in part drowned by the sea. These valleys are thought to have formed after the glacier retreated away from the upstream end of the outwash plain and after the outwash plain surface became inactive. All of the outwash plains on the Cape and Islands are incomplete. They have had their downstream ends eroded away by the sea and many others have had their upstream ends and ice-contact heads partly or completely removed in a similar manner. Some had their upstream ends destroyed by readvances of the glacier during the formation of an end moraine.

Kame and Kettle Terrains

In some places on the Cape and Islands, the land surface consists of hills that are conical or have steep sides and a relatively flat top, called kames, and closed depressions, called kettles (Fig. 34). The kames formed when sediment filled holes in the glacier. The surface is underlain by meltwater stream deposits that are collapsed to the extent that the original, gently sloping, stream-graded surface of the deposit is completely or almost completely destroyed. This kame and kettle terrain is underlain by ice-contact deposits or by outwash. If the ice blocks were only partly buried, the surface may be boulder strewn in a manner similar to an ice-contact head. Kame and kettle morphology develops where broad areas of stratified drift are deposited atop ice that is mostly thicker than the overlying deposits. When the ice melts, all or all but minor parts of the alluvial surface collapse.

Kettles

Although kettle holes are briefly described above, they are a major feature of the glacial landscape and require additional discussion. They were formed when ice blocks left behind by the retreating ice were buried by glacial drift, mostly outwash.

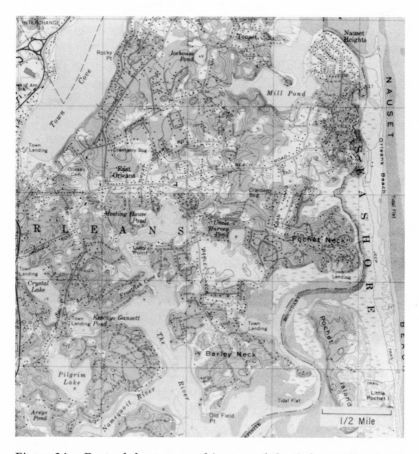

Figure 34.   Part of the topographic map of the Orleans 7½-minute quadrangle (original scale 1:24,000) showing kame and kettle terrain. Glacial sediments were deposited over ice by meltwater streams. When the buried ice melted away, the deposits collapsed so that most if not all of the alluvial surface was destroyed.

As the ice melted, the deposits atop the ice collapsed to form a depression. The largest kettles occur in the outwash plains on upper Cape Cod. Their complex shape suggests that they were probably formed by the melting of two or more closely spaced ice blocks. The size of the ice block assemblage can be estimated by the size of the kettle, and the largest were probably close to 1,000 acres in area. An indication of the minimum thickness of the ice block can be gained from the depth of the kettle, if we assume that the ice was completely buried. In most places, this assumption is probably correct because only a few kettle holes show an indication of material added to the upper part of the outwash deposits by the ice block. The depth is measured from the restored precollapse outwash plain surface, determined by the maximum altitude of the outwash plain adjacent to the kettle hole. For example, the restored altitude of the outwash plain adjacent to Cliff Pond kettle in Brewster is about 140 feet above sea level, and the floor of the kettle has an altitude of about 60 feet below sea level. Thus, the kettle depth indicates that the ice block was about 200 feet thick. This estimate is reinforced by the fact that glacial ice less than 200 feet thick is no longer plastic, but brittle. We do not know how long the buried ice persisted. It is possible that deeply buried ice blocks were well insulated from the warmer temperatures of the postglacial climate and may have taken several thousand years to melt away. This estimate is supported by the oldest radiocarbon date on vegetation from the bottom of a kettle pond in Wellfleet. The age of the vegetation indicates that the pond formed about 12,000 years ago and at least several thousand years after the retreat of the Laurentide ice. A less reliable, but more intriguing, indication of the persistence of ice buried by outwash comes from Indian folklore. The story tells of how the Indians observed the formation of Walden Pond, a kettle pond in Concord, Massachusetts. If the story is true, this event could not have occurred more than about 10,000 years ago.

Kettle Ponds

Kettle ponds (a few are called lakes) are a distinctive feature of the glaciated landscape. They are so common to outwash plains, in the areas north of the glacial limit, that they allow recognition of outwash plains on maps without contours, on aerial photographs, or from airliners and satellites flying high over the Cape and Islands (Fig. 35). Kettle ponds form where the bottom of a kettle is below the ground-water table. They range in size from a few tens of feet across to hundreds of acres. The largest kettle lake is Long Pond in Brewster and Harwich, which covers 743 acres. Water depths are equally varied and range from less than 10 feet to more than 80 feet. At about 85 feet, Mashpee Pond in Mashpee and Cliff Pond in Brewster are the deepest.

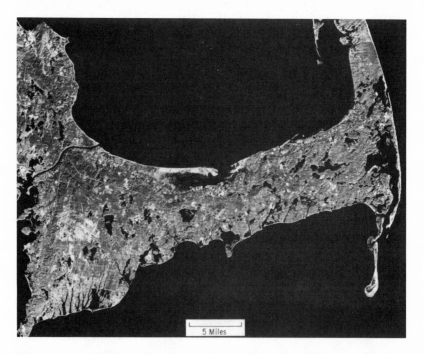

Figure 35. Satellite photograph of upper Cape Cod, showing the numerous kettle ponds in the outwash plains. The large embayments in Chatham and Orleans were formed by sublobes of the South Channel lobe that excluded outwash deposition in these places. Valleys cut into the Mashpee pitted plain, in the lower left of the photograph, have been drowned by sea-level rise to form fingerlike embayments.

The shores of kettle ponds are influenced by waves and currents in the same manner as the seashore. Cliffed headlands occur along the shore, cut by wave attack (Fig. 36). The material eroded from the cliff is transported by longshore currents and longshore drift to where it is redeposited to form beaches, baymouth bars across small coves, and the nearshore sandy bottoms. These processes tend to smooth the shoreline. Some kettle ponds have become almost round (Fig. 37). Finer grained debris, including abundant organic matter, is deposited in the deeper parts of the pond. This organic mud has been cored and dated. Pollen and radiocarbon ages from Duck Pond (Fig. 36) in Wellfleet provide a 12,000-year record of the changes in vegetation in and adjacent to the pond.

Many kettle ponds have no surface inlet or outlet, but they are not stagnant, because they act like windows into the ground water that flows through the glacial deposit towards the sea much like a stream, but at a much slower rate. The slowly moving ground water constantly refreshes the water in the pond.

Valleys Cutting Outwash Plains

The outwash plain surfaces are cut by clearly defined valleys in many places (Fig. 38). On lower Cape Cod, the valleys are called pamets, after the Pamet River valley in Truro, or hollows; on Martha's Vineyard, they are called bottoms. The valleys tend to be straight, roughly parallel to each other, and generally spaced less than a mile apart. They have flat floors and steep sides. The valleys may have short tributary valleys, but lack the dendritic or treelike pattern common to most valley systems. In places, the valleys are interrupted by kettles and many have kettles at their head. Along most of the coast, the lower reaches of the valleys have been destroyed by marine erosion. However, on Martha's

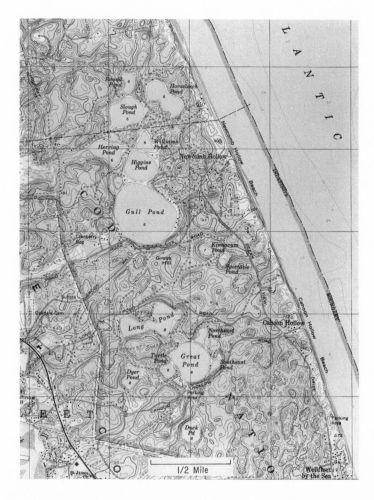

Figure 36. Part of the topographic map of the Wellfleet 7½-minute quadrangle (original scale 1:25,000), showing kettle ponds in the Wellfleet outwash plain. The ponds form when the kettles are deep enough to intersect the water table. The elevations of the pond surfaces increase slightly from north to south, indicating that the ground water surface slopes northward in this region. The closely spaced ponds in the northern part of the map may have formed over a single ice block. Gull, Higgins, and Williams Ponds are now separated by small spits formed of sand eroded from cliffed sections of the shores.

Figure 37. Aerial photograph of Great Pond, a kettle in the Wellfleet outwash plain. Baymouth bars have formed across irregularities in the kettle hole so that the pond shore is nearly circular. The processes that have smoothed the pond shoreline are similar to those that form spits along the ocean shore. Great Pond is roughly 1,500 feet in diameter. Photograph by the Cape Cod National Seashore.

Vineyard, they appear to be tributary to larger valleys that are now occupied by large coastal ponds. However, almost all of the tributaries occur along the northeast side of the trunk valley. Narrow embayments, tidal creeks and marshes occupy the lower reaches of the outwash plain valleys. Good examples of the valleys can be seen in Wellfleet (Fig. 39, upper) and Falmouth (Fig. 39, lower) on Cape Cod and on Martha's Vineyard.

The valleys are an enigma. They are clearly not being cut today, because most lack modern streams. This is a result of a sandy soil and highly permeable outwash, which encourages perco-

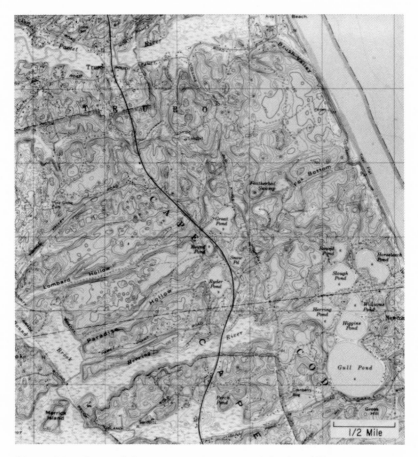

Figure 38.   Part of the topographic map of the Wellfleet 7½-minute quadrangle (original scale 1:25,000), showing dry valleys, including the Pamet Valley at the top of the map. In places, the valleys are interrupted by kettles, indicating that the valleys formed before the ice blocks in the Wellfleet outwash plain melted away.

lation and prevents runoff during rainstorms or during rapid snow melt between winter storms. Many are interrupted by kettles, which shows that they were cut before the ice blocks buried beneath the outwash melted, an event that must have occurred within a few thousand years following deglaciation.

Figure 39. A view down the Pamet Valley in the Wellfleet outwash plain (upper photograph) and down the Connamessett Valley (lower photograph) in the Mashpee outwash plain. The Pamet Valley is the largest valley on Cape Cod; it completely crosses the lower Cape, and the drift surface beneath the saltwater and freshwater peat is well below sea level throughout. It is separated from the open ocean by a baymouth bar at the east end. The Connamessett Valley is more typical of the outwash plain valleys that are commonly the sites of cranberry bogs. Connamessett Valley photograph by Dann S. Blackwood of the U.S. Geological Survey.

It is not fully clear whether they are glacial or postglacial features. Many investigators of the glacial geology of the Cape and Islands have written that the valleys were cut by meltwater streams. However, the valleys stop short of the upper parts of the outwash plains where the meltwater source would have been located. In addition, they probably would not have formed so long as braided meltwater streams carrying abundant outwash continually swept across the outwash plain, filling developing kettle holes and any other depression in the outwash plain surface. Thus, it appears that they were cut after the outwash plain surface became inactive. The valleys do not resemble stream-cut valleys. They do not form a dendritic drainage pattern that is typical of a gently sloping surface undergoing dissection by surface streams. If they were cut by streams fed by rain and snow melt, then conditions other than those that exist today must have been present because now only a few streams occur on the Cape and Islands, and those are not fed by runoff. Permafrost, as a layer of permanently frozen ground in the upper part of the outwash plain deposits, may have stopped surface water from percolating into the ground and thus may have fostered runoff. However, fossil frost wedges, that are generally accepted as proof of past permafrost, are not found in the glacial drift. The late-glacial climate was probably too warm for permafrost to form.

For some time after the outwash plains were formed, they were barren or sparsely vegetated. Without vegetation and ground litter, the ubiquitous silty eolian layer may have caused greater runoff. Today, organic litter in the upper part of the modern soil traps water, discourages runoff, and encourages percolation.

The valleys may not have been cut by surface water at all. The asymmetrical tributary pattern of the valleys in the Martha's Vineyard outwash plain may provide a clue to their origin. These tributary valleys lie on the side of the trunk valley that is farthest up the water table. This suggests that the valleys may have been formed by spring sapping, a process by which springs of ground water migrate up a sloping surface. The water they discharge forms steep-headed channels. A small-scale example of spring sapping can be seen today on the ocean beaches as ground water reaches the sloping face of the beach, forming a spring, which then migrates up the beach slope by eroding small, shallow channels or beach rills in the beach sand. In most places, however, springs do not presently emerge on the outwash plains because the water table lies below the surface. The water table must have been higher for spring sapping to work. The answer may lie with the glacial lakes that developed between the retreating ice front and the outwash plain deposits, as one did in Cape Cod Bay. For example, the initial stages of the Cape Cod Bay lake were 80 and 60 feet above sea level, so they would have supported a much higher water table beneath inner Cape Cod. The Sandwich moraine and the outwash plains may have functioned as leaky dams with springs that emerged from the outwash plain some distance south of the present shoreline of Nantucket Sound. These springs may have then migrated upslope by spring sapping, and their migration may have formed the valleys. When the level of the glacial lake in Cape Cod Bay fell to below present sea level, the water table fell and spring sapping ceased. Laboratory experiments using these conditions have produced channels very much like the outwash plain valleys. Until more information is available, permafrost or lack of vegetation to account for the enhanced runoff and spring sapping are reasonable hypotheses.

Some of the larger channels on Cape Cod that have been eroded into the drift may have a different origin. The deep channel now occupied by the Cape Cod Canal, Bass River between Yarmouth and Dennis, and Town Cove in Orleans are considered to have been outlets of glacial lake Cape Cod. These channels and the Pamet Valley may have been, in part, cut during catastrophic draining of glacial lakes. Groups of very large boulders encountered during the construction of the canal could be considered evidence of a major catastrophic flow from glacial lake Cape Cod. The Bass River valley and the low that includes Town Cove provide nearly sea level passages across Cape Cod and are clearly different from the dry outwash plain valleys and must have had a different origin, one that may have included catastrophic outflows. The Pamet River valley is larger by far than the other valleys in the Wellfleet outwash plain and may have been carved by floods draining a lake that existed east of the lower Cape. It is easy to imagine that the Pamet was the longest of the spring sapping valleys in the Wellfleet plain and that it eventually breached the dam for a lake east of lower Cape Cod. Once breached, the unleashed flood waters would have widened and deepened the Pamet River valley. Although all this is speculative, and good evidence is lacking, it is likely that at times during the formation of glacial Cape Cod, catastrophic outflows from beneath the ice or from glacial lakes dammed by drift did occur. Catastrophic outflows are so common in glacially covered parts of Iceland that they have a name, jökulhlaups, and such massive sudden outflows of glacial lakes are thought to have created many of the landforms around Spokane, Washington.

GLACIAL LAKES

Glacial lakes were a large aspect of the landscape during the retreat of the Laurentide ice from the Cape and Islands. The lakes were temporary features that owed their existence to the glacier that dammed free drainage from low areas such as the sounds and the major bays. The best known glacial lake occupied Cape Cod Bay and is called Glacial Lake Cape Cod. It was dammed to the north by the Cape Cod Bay lobe and to the east by the South Channel lobe. Initially the lake was small, but as the Cape Cod Bay lobe retreated northward, the lake greatly increased in size. Evidence for glacial Lake Cape Cod includes outwash deltas along the lake shore, silt and clay beneath the outwash deposits, and glacial lake deposits around Cape Cod Bay. Evidence for a glacial lake in Nantucket and Vineyard Sounds comes from a deep borehole in Harwich that penetrated more than 100 feet of glacial lake clay beneath the Harwich outwash plain deposits. In addition lake deposits are inferred from offshore seismic profiles that show reflectors from horizontally laminated lakelike sediments. There also may have been glacial lakes in Buzzards Bay, in the Gulf of Maine, and in Great South Channel. The lake sediments in the Highland plain deposits at Highland Light in Truro may be a remnant of deposits formed in a large lake east of Cape Cod. The glacial lakes were short lived and drained as the damming ice retreated out of the basins or when dams of glacial drift eroded. From time to time, some of the lakes probably drained catastrophically when the ice or drift dams burst.

GLACIAL BOULDERS

Boulders are a common feature on the surface of the moraine and ice-contact deposits. Large to very large boulders provide direct evidence of glaciation because they are too big to have been carried by meltwater streams. Other boulders show evidence of glacial transport because they are soled (have a flat, striated, glacially ground surface) (Fig. 40). Many of these boulders were transported scores of miles by the glacier before being deposited on the Cape and Islands. The larger boulders measure 25 or more feet long. The largest known boulder is Enos or Doane Rock in Eastham (Fig. 41). This very large boulder stands 15 feet above ground level and is about half buried. It measures 40 feet long and 25 feet wide. It is composed of a volcanic rock that is common in the Boston area, and it may have come from there. However, it is more likely to have been plucked from the bedrock beneath the Gulf of Maine because it was deposited from ice that came from that direction. Although Doane Rock is considered to be the largest glacial boulder on Cape Cod, there may have been larger ones. Many of the very large boulders have broken into fragments as freezing water and tree roots force the blocks apart along natural planes of weakness called joints. These very large boulders were deposited directly by the ice, either along the front of the glacier or from ice blocks left behind by the retreating glacier. Even into the early years of this century, many large glacial boulders were quarried for building stone, and many areas of the Cape and Islands are less stony than they were when the glacier retreated because of this

Figure 40. Soled and striated basalt boulder from the Nauset Heights deposits. This boulder was carried in the base of the ice and was ground flat as it was dragged across the underlying bedrock. Similar striations on bedrock surfaces indicate the direction of glacial flow, which in southern New England was generally southeastward.

industry. One type of building stone, called West Falmouth Pink Granite, was extensively quarried from boulders in the Buzzards Bay Moraine around that village. The West Falmouth granite is most likely a variety of the Dedham Granodiorite that is the dominant bedrock on the western shore of Buzzards Bay. This pink granite was chosen for President Kennedy's memorial in Arlington National Cemetery. In the past, on the Cape and Islands, many smaller boulders, up to a few feet in diameter, were used to build stone walls. Today, some of the areas where boulders abound are mined to provide boulders for shore structures such as groins, jetties, and sea walls.

Figure 41.   Doane Rock, the largest glacial boulder on Cape Cod, is located on the Eastham outwash plain. The boulder is half buried by outwash and wind-blown sand. Many boulders are too large to have been carried by running water and must have melted out of partly buried ice blocks or from the glacier itself. Photograph by Dann S. Blackwood of the U.S. Geological Survey.

STONES AND OTHER BITS AND PIECES IN THE DRIFT

Most, if not all, of the different rock types that make up the bedrock in southeastern Massachusetts are represented by stones found in the drift of the Cape and Islands. The most common stones in the drift include granite (by far the most abundant), volcanic rocks, basalt, and quartzite (in decreasing abundance). These proportions probably reflect the relative amounts of these rock types in the bedrock north of and beneath the Cape and Islands region. A few stones found on the Cape and Islands can be traced northward to very restricted source areas. These stones are called indicator stones. Fragments of dark-gray silicified wood are found in the drift of western Cape Cod and Martha's Vineyard. They can be traced back to a bedrock source near Middleboro, Massachusetts. Pebbles and cobbles of a gray to white quartzite that contain fossils of a Late Cambrian bivalve, called a brachiopod, can be found throughout the Cape and Islands. The only know source for these pebbles and cobbles is a Carboniferous conglomerate (a consolidated sand and gravel) near Dighton, Massachusetts. The source for the fossiliferous stones now found in the conglomerate has never been located. Stones of red felsite, a very fine-grained volcanic rock, are also found throughout the Cape and Islands. The only known bedrock source is just south of Boston.

The areas where these rock types are found in the glacial drift are fan-shaped, and the apex of the fan is at the source outcrop. Several indicator stones are found in the glacial drift in westernmost Martha's Vineyard and the southwesternmost part of the Elizabeth Islands. One indicator stone, a schist containing large cross-shaped crystals, called chiastolite comes from east-central Massachusetts, a distance of more than 80 miles. Two other indicator stones come from northeastern Rhode Island. One is an agate, a variety of quartz, and the other is an iron-rich, coarse-grained igneous rock called cumberlandite. The distribution of a unique rock type in the drift is called a train and provides information on the flow direction of the Laurentide ice, which in southeastern New England was southeasterly. However, the occurrence of similar rock types in deposits of more than one glacial lobe may indicate other source areas not yet discovered.

Other stones and fossils with no known source show what other kinds of rocks and strata were also overridden by the Laurentide ice. Cobbles and pebbles of red sandstone are relatively abundant in the drift in Orleans on Cape Cod. The rock is similar to the Triassic red beds in the Connecticut Valley, so these stones imply that Triassic rocks lie beneath parts of lower Cape Cod and the Gulf of Maine. Such Triassic deposits are now known from offshore seismic profiles and other geophysical studies. Pieces of a shelly sandstone of Eocene age and sharks' teeth (Fig. 42) found in the drift of outer Cape Cod indicate that marine

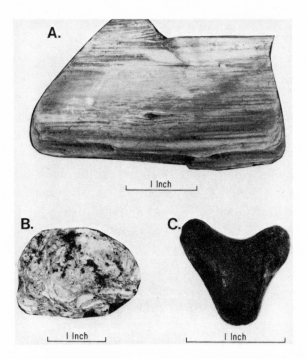

Figure 42.   Silicified wood (A), shelly marl (B), and shark tooth (C) from the Wellfleet plain deposits. The silicified wood may be Cretaceous or Tertiary in age. The shelly marl, a limestone, and the shark tooth are Tertiary in age. These materials indicate that the glacier advanced over coastal plain and continental shelf deposits as well as bedrock on its way to the Cape and Islands.

strata of Tertiary age occur in the Gulf of Maine. Fragments of a light-colored silicified wood of Cretaceous or Tertiary age provide sparse evidence that pre-Pleistocene terrestrial strata are also buried somewhere beneath the glacial drift.

Pieces of wood (Fig. 43) and shells picked up during the advance of the last ice and incorporated into the glacial drift probably represent the Sangamonian interglacial stage and the early to

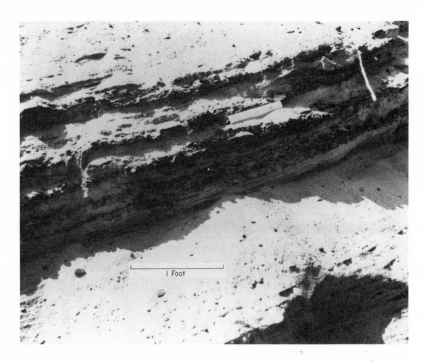

Figure 43. *Wood fragments in the Wellfleet plain deposits. Although these wood fragments were not carbon-14 dated, other wood and shells from the Wellfleet plain deposits provided carbon-14 ages that are considered unreliable because of contamination by younger organic matter or were too old to date. Wood and shell in the glacial drift indicate that Laurentide ice advanced over a landscape that was similar to coastal southern New England today.*

middle part of the Wisconsinan stage, when the Laurentide ice was still north of southern New England. The wood fragments and shells have provided numerous radiocarbon ages that establish the late Wisconsinan age of the drift. The youngest of these ages, on the order of 21,000 years, may roughly approximate the time when the Laurentide ice reached the Cape and Islands. The wood and shell also indicate that the Laurentide ice sheet overrode forests and sea floor, a landscape that may have been similar to southeastern Massachusetts today.

LATE WISCONSINAN HISTORY

The late Wisconsinan glacial deposits, and their corresponding landforms on the Cape and Islands, are not all the same age. Their relative ages as determined by geologic mapping provide a history of events that occurred during the retreat of the Laurentide ice from the region. Direct correlations between the glacial deposits on Martha's Vineyard, Nantucket, and Cape Cod cannot be made because they are physically separated. However, because the glacial deposits on the islands lie close to the maximum advance of the Laurentide ice (Fig. 20), they are clearly older than those that make up Cape Cod, which could only have been deposited after the Laurentide ice had retreated from the islands.

The Pleistocene geology of Martha's Vineyard, including Nomans Land, was most recently mapped by Clifford A. Kaye

(Fig. 44).[1] Kaye recognized deposits of three moraines—the Squibnocket, Gay Head, and Martha's Vineyard moraines (Qs, Qgh, and Qmv). He considered the Squibnocket and Gay Head moraines to be of Illinoian age. The youngest moraine (Qmv) he considered to be early Wisconsinan in age. In the eastern part of the island, this moraine is overlain by stratified drift that forms an outwash plain (Qmvo), which was considered by Kaye to be of early Wisconsinan age also.

A previous geologic map of Martha's Vineyard and Nomans Land, by Edward Wigglesworth, was published in 1934. This map showed only a single moraine, made up of the three moraines of Kaye, and a single outwash plain, both of which were assigned a Wisconsinan age.

Although Kaye's map is included in this book and was used to compile the USGS satellite geologic map of the Cape and Islands (I-1763, see Appendix C), I believe that his interpretation should be modified. I suggest that indeed there is only one moraine and that it is confined to the northwestern part of the island west of Vineyard Haven, although in the eastern part of the island it may occur beneath outwash deposits. The glacial drift east of Vineyard Haven forms a single outwash plain. The outwash plain has a collapsed head, which runs from Vineyard Haven to Edgartown. The difference between the western and eastern parts of the island can be seen clearly when observed from the Falmouth shore of Vineyard Sound or from the deck of a ferry approaching Vineyard Haven. To the west of Vineyard Haven,

---

[1] *The letter symbols on the geologic maps (Figs. 44 through 47) are used to identify the geologic map units. Their relative ages are shown by their positions vertically in the map explanation in Figure 44; age increases from top to bottom. In Figures 45 and 46, the relative age of the map units is shown by their position vertically in the correlation of map units. Geologic units that are considered to be contemporaneous have the same vertical position.*

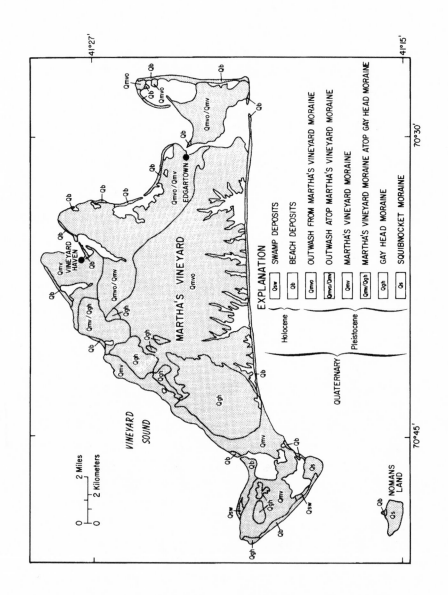

84

the moraine produces a skyline that is higher and irregular, while, to the east of Vineyard Haven, the outwash plain forms a skyline that is lower and flat. I believe both the moraine and outwash plain are late Wisconsinan in age. The moraine was formed first, by an advance of the Buzzards Bay lobe. The outwash plain was formed somewhat later, by meltwater from the Cape Cod Bay lobe during the initial stages of retreat.

On Nantucket, the glacial drift makes up a drumlin (a stream-lined hill formed beneath the glacier), a single moraine, and four outwash units. The oldest late Wisconsinan deposit on Nantucket is a till that underlies the low drumlin, about 3 miles southwest of Nantucket Village (Qwd, Fig. 45). The drumlin and underlying till were probably formed when a sublobe of the Cape Cod Bay lobe advanced some distance south of the island. The outwash deposits of the older Nantucket plain (Qno_1) were laid down next. A lack of kettles in that outwash plain and the existence of younger kettled outwash plain deposits west and east of the older Nantucket outwash plain indicate that the older outwash was deposited in an interlobate angle between a sublobe of the Cape Cod Bay lobe and the South Channel lobe. The southeast trend of the ice-contact slope on the east side of the older Nantucket plain further supports this conclusion.

The Nantucket moraine (Qnm) formed next, during an advance of the Cape Cod Bay lobe that in part overrode the ice-contact head of the older plain. The southeast-trending part of the ice-contact head of the older plain and the Nantucket moraine are in part buried by outwash of the Siasconset plain

Figure 44. (opposite)   Geologic map of Martha's Vineyard. The relative ages of the geologic units is shown by the explanation. From a 1:31,680 map compiled by Clifford A. Kaye of the U.S. Geological Survey.

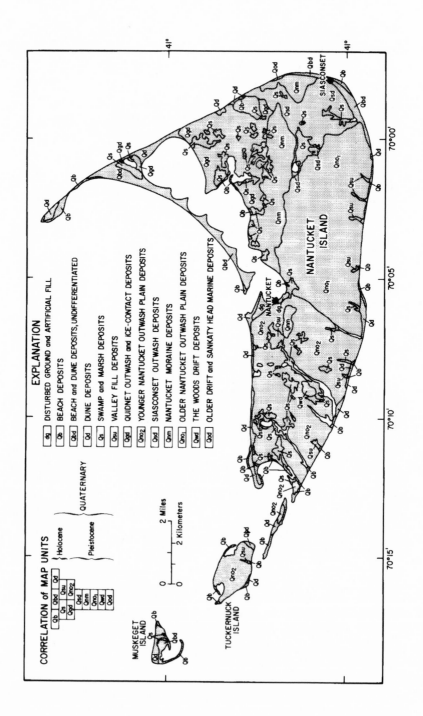

CORRELATION of MAP UNITS

Qb	Qbd	Qd		
Qs		Qsu	Qno₂	
Qgd				
		Qsd		
		Qnm		
		Qno₁		
		Qwd		
		Qod		

Holocene ⎫
          ⎬ QUATERNARY
Pleistocene ⎭

EXPLANATION

- **dg** DISTURBED GROUND and ARTIFICIAL FILL
- **Qb** BEACH DEPOSITS
- **Qbd** BEACH and DUNE DEPOSITS, UNDIFFERENTIATED
- **Qd** DUNE DEPOSITS
- **Qs** SWAMP and MARSH DEPOSITS
- **Qsu** VALLEY FILL DEPOSITS
- **Qgd** QUIDNET OUTWASH and ICE-CONTACT DEPOSITS
- **Qno₂** YOUNGER NANTUCKET OUTWASH PLAIN DEPOSITS
- **Qsd** SIASCONSET OUTWASH DEPOSITS
- **Qnm** NANTUCKET MORAINE DEPOSITS
- **Qno₁** OLDER NANTUCKET OUTWASH PLAIN DEPOSITS
- **Qwd** THE WOODS DRIFT DEPOSITS
- **Qod** OLDER DRIFT and SANKATY HEAD MARINE DEPOSITS

NANTUCKET ISLAND

MUSKEGET ISLAND

TUCKERNUCK ISLAND

SIASCONSET

NANTUCKET

0   2 Miles
0   2 Kilometers

41°   41°

70°15'   70°10'   70°05'   70°00'

86

(Qsd), which was formed when the ice front had retreated to a point north of the moraine and meltwater streams drained through the low in the moraine of Siasconset village.

Continued retreat of the Cape Cod Bay lobe placed the ice front north of Nantucket while the two youngest drift units were deposited. The greatly collapsed Quidnet outwash and ice-contact deposits (Qgd) were laid down in the eastern part of the island, probably in an ice-choked lake that may have received sediment from both the Cape Cod Bay and Great South Channel lobes. The younger Nantucket plain outwash (Qno2), to the west of Nantucket Village, was deposited over numerous ice blocks left behind during the retreat of the western sublobe from its maximum advance south of the island. The relative ages of these two deposits cannot be established because they are not adjacent to one another, however, they are clearly the youngest glacial drift units on Nantucket and could have been contemporaneous.

Cape Cod is several times larger than Martha's Vineyard or Nantucket and consists of many late Wisconsinan glacial drift units. These units preserve the record of the three lobes following the retreat of the Laurentide ice sheet from the islands.

The oldest upper Cape glacial deposits form a number of isolated hills along the Nantucket Sound shore (Fig. 46, 47). The easternmost is Great Hill in Chatham and the westernmost is Falmouth Heights. These features were mapped as the Nantucket Sound ice-contact deposits (Qnd). They were probably deposited in holes or reentrants in the ice when the ice front of the Cape Cod Bay lobe was a short distance south of the south

Figure 45. (opposite)   Geologic map of Nantucket, Tuckernuck, and Muskeget Islands. The relative ages of the geologic units are shown by the correlation of map units. Modified from a 1:48,000 scale map (I-1580) published in 1985 by the U.S. Geological Survey.

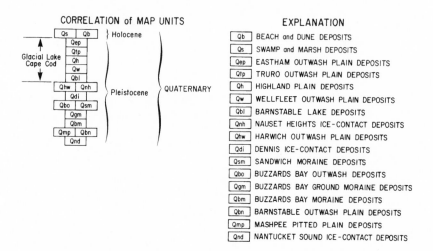

Figure 46. Correlation of map units and explanation for the geologic map of Cape Cod.

shore of the upper Cape. These ice-contact deposits are in part overlain by younger outwash, including deposits of the Mashpee, Barnstable, and Harwich outwash plains (Qmp, Qbn, and Qhw). The deposits of the Mashpee outwash plain contain a mixture of stones indicating that the source of meltwater came from both the Buzzards Bay lobe and the Cape Cod Bay lobe. The Barnstable outwash plain deposits contain stones indicating that the meltwater source was solely the Cape Cod Bay lobe. Both these outwash plains were formed after the Buzzards Bay and Cape Cod Bay lobes had retreated to positions approximated by the present west and north shores of the upper Cape. Readvance of the Buzzards Bay lobe deformed the Mashpee outwash plain deposits and the glacial drift in Vineyard Sound to form the Buzzards Bay moraine (Qbm). As the Buzzards Bay lobe retreated away from the moraine, it deposited ground moraine (Qgm), a till and ice-contact deposit with a boulder-strewn, hummocky surface of low relief, to the west of the end moraine. Shortly afterwards, the Cape Cod Bay lobe

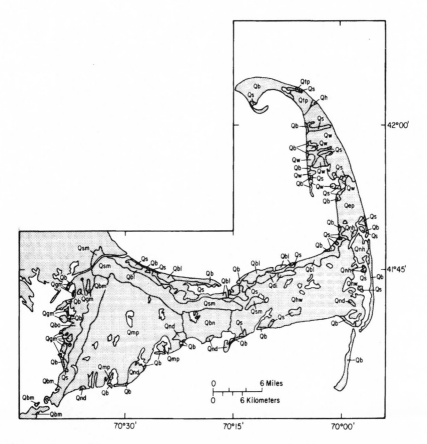

Figure 47.   Geologic map of Cape Cod. Modified from a 1:1,000,000 scale map (I-1763) published in 1986 by the U.S. Geological Survey

advanced against the Mashpee plain and Barnstable plain deposits to form the Sandwich moraine (Qsm). Meltwater draining the Cape Cod Bay lobe west of the Sandwich moraine deposited outwash (Qbo) atop the ground moraine.

Following the formation of the Sandwich moraine, meltwater deposits from the Cape Cod Bay lobe formed the kames of the Dennis ice-contact deposits (Qdi) and, a short time later, the Harwich outwash plain (Qhw).

As the Cape Cod Bay lobe subsequently retreated northward from the Sandwich moraine and the Harwich plain, the earliest stages of a glacial lake developed beyond the ice front. Lake-floor deposits of this lake and ice-contact deltas (Qbl) occur along the north flank of the Sandwich moraine and the ice-contact head of the Harwich outwash plain and are the youngest deposits of the Cape Cod Bay ice lobe on Cape Cod. However, they are not the last record of the retreat of this lobe. Billingsgate Shoal off Wellfleet and Eastham is underlain by a moraine formed during a minor readvance of the Cape Cod Bay lobe. This moraine and a series of small interlobate moraines north of the Cape Cod Canal provide a record of the retreat of the Cape Cod Bay and Buzzards Bay lobes after the formation of the glacial part of upper Cape Cod.

The proglacial lake that developed between the upper Cape and the retreating ice front is formally called Glacial Lake Cape Cod. The lake was a major factor in the formation of most of the drift on lower Cape Cod. Westward-flowing meltwater streams across the lower Cape drained the South Channel lobe and entered various stages of the lake. The lake-level stages were determined by the altitude of various lake outlets, which were sequentially lower. Initially, the lake drained through the Sandwich moraine by way of a topographic low between Cape Cod and Buzzards Bays. The early stages of Glacial Lake Cape Cod were on the order of 80 and 60 feet above present sea level. The ice-contact deltas along the south shore of Cape Cod Bay provide evidence of these early lake-level stages. The oldest glacial drift unit on lower Cape Cod is a outwash delta. The levels of this delta and of corresponding outwash deltas on the west side of Cape Cod Bay were controlled by the outlet between Cape Cod Bay and Buzzards Bay when the lake level had dropped to about 50 feet above present sea level. This outlet threshold was continuously lowered by erosion until it reached an altitude of about 30 feet (the altitude of the divide between the Scussett and

Rivers before the canal was built) and was abandoned for a lower outlet elsewhere. The Scussett and Manomet Valleys formed a natural route for a portage between Cape Cod and Buzzards Bays during Indian and colonial times. Much later, the valleys became the site of the Cape Cod Canal, which relieved many ships of the hazards of the shoals in Nantucket Sound and Nantucket Shoals south of the island.

Later levels of Glacial Lake Cape Cod must have been controlled by outlets lower than 30 feet above sea level. The locations of these lower spillways are not clear. One may have been through Town Cove in Orleans and Eastham. Much of the lake-level change probably occurred gradually, but from time to time the lake level may have fallen very quickly as catastrophic floods broke through the dam formed by the glacial drift.

The Nauset Heights ice-contact deposits (Qnh) are thought to be roughly equivalent in age to the Harwich outwash plain deposits and to be the oldest glacial drift unit on lower Cape Cod. The deposit was formed between two sublobes by meltwater streams draining the South Channel lobe. The sublobes occupied the present site of Pleasant Bay in Chatham and Orleans and Nauset Harbor in Orleans and Eastham. Meltwater streams draining west from the South Channel lobe formed the Wellfleet plain (Qw), which is the largest and highest outwash plain on lower Cape Cod. The altitude of the contact between the foreset and topset beds in the Wellfleet plain deposits, about 50 feet above present sea level, shows that the outwash plain is a delta that was controlled by a stage of Glacial Lake Cape Cod at a similar level. At that time, the lake had greatly increased in size because the Cape Cod Bay lobe had retreated to a position north of Wellfleet. An ice-contact slope forms the contact between the Wellfleet plain and the Eastham plain to the south and indicates that the deposits of the Wellfleet plain were laid down against the Nauset Harbor sublobe.

The northern limit of the plain is also an ice-contact slope, which makes up the contact between the Wellfleet plain deposits and the deposits of the Highland plain and the Truro plain. This ice-contact slope marks the interlobate angle between the Cape Cod Bay and South Channel lobes. The angle was located a bit south of Highland Light in Truro.

Deposits of the Highland plain (Qh) were laid down in a lake, independent of Glacial Lake Cape Cod, which formed following a retreat of the interlobate angle away from the Wellfleet plain ice-contact slope. The Highland plain deposit may be a small remnant of deposits, now eroded, that were formed in a large glacial lake that developed between the retreating South Channel lobe and the Wellfleet plain.

The interlobate angle again migrated north and east as the Cape Cod Bay and South Channel lobes retreated. Meltwater streams from the South Channel lobe again entered Glacial Lake Cape Cod to deposit the Truro plain (Qtp).

In the southern part of lower Cape Cod, the Eastham plain (Qep) was formed when the Nauset Harbor sublobe retreated away from the Nauset Heights deposits and the ice-contact slope of the Wellfleet plain. The Eastham plain outwash deposits were laid down in the area freed of ice. Exposures of the Eastham plain deposits show them to be fluvial. However, boreholes show that the outwash overlies lake deposits, and thus it is likely that the Eastham plain is a delta controlled by a very low level of Glacial Lake Cape Cod.

With the deposition of the Eastham plain, the glacial development of Cape Cod ceased. The Buzzards Bay lobe had by then retreated to the north of Buzzards Bay, and the Cape Cod Bay lobe had retreated into Massachusetts Bay. Continued retreat of the South Channel lobe away from the east side of the Cape and into the Gulf of Maine caused Glacial Lake Cape Cod to drain. The retreating lobe also temporarily allowed the sea to enter the northeastern part of Cape Cod Bay (Fig. 48). This late Wisconsinan marine incursion was caused by the depression of the Earth's crust under the weight of the overlying glacier.

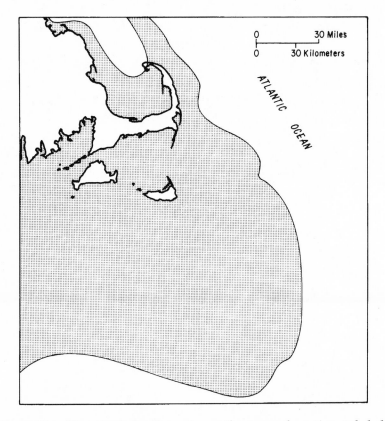

Figure 48.   The stippled pattern shows the exposed continental shelf during the early stages of the retreat of the Laurentide ice. The Atlantic Ocean was close to the lower Cape, and even entered the northeastern part of Cape Cod Bay because of the deep basins in the Gulf of Maine and because the bedrock surface was depressed below the late Wisconsinan low sea-level stand by the weight of the overlying ice. The continental shelf south of the Islands was beyond the glacial limit and remained above the late Wisconsinan low sea-level stand.

However, this incursion of the sea was short lived because, following the retreat of the ice, the crust rebounded rapidly.

The sequence of glacial deposits on the Cape and Islands indicates the order of retreat of the lobes. In general, the Buzzards Bay lobe deposits are the oldest and the South Channel lobe deposits the youngest. Thus, the retreat of the lobes progressed from west to east. It did not take long for the Laurentide ice sheet to retreat from the region. In the Gulf of Maine, new radiocarbon ages from a glaciomarine clay that was deposited beyond the ice front show that Laurentide ice had retreated north of Cape Cod shortly after 18,000 years before present. If the estimate for the start of deglaciation is about right, then these dates from the Gulf of Maine indicate that retreat from the Cape and Islands took only a few thousand years at most. This short time period indicates just how dynamic the glacial environment was. Moraines and outwash plains must have been formed in a few hundred years or so and all of the glacial Cape and Islands were formed in a few thousand years.

SOUTH OF THE LAURENTIDE ICE SHEET

When the Laurentide ice sheet had retreated north of Cape Cod, the glacial development of the region ended, and the Cape and Islands waited for the next major chapter in their geologic history, the transgression of the sea. At the height of the Wisconsinan glaciation, worldwide sea level was approximately 300 feet below its present level, and the shoreline was 75 miles south of the Islands. To the east, the shore was much closer because the nearby deep basins in the Gulf of Maine were submerged early. Lower sea levels during the Wisconsinan glaciation exposed broad areas of the continental shelf beyond the limit of the Laurentide ice sheet (Fig. 49). Although the evidence is sparse, because of the late Wisconsinan and Holocene marine transgression and the submergence of the shelf, patches of peat and bones and teeth

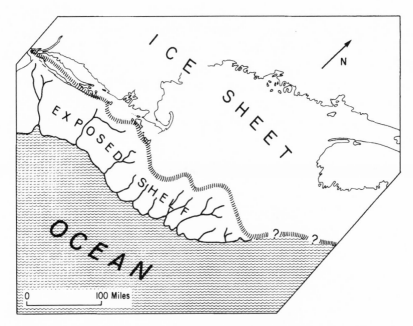

Figure 49.   The exposed continental shelf during the maximum advance of the Laurentide ice. Streams crossing the shelf were probably largely fed by meltwater from the glacier. Conditions on the shelf were probably harsh, but it provided a refuge for many of the plants and animals that migrated northward and colonized the glaciated landscape as the ice retreated.

dredged up from the sea floor indicate the presence of plants and animals that lived on the exposed shelf south of the ice sheet. The climate on the exposed shelf was undoubtedly more severe than exists in the Cape and Islands today, but it was certainly temperate enough to support abundant and diverse plant and animal populations. Many of these plants and animals would be familiar today in the northeastern United States and eastern Canada. Forests were made up of spruce, fir, and pine trees, with open parklands made up of tundra plants and grasses. Scattered about were abundant swamps and marshes on the gently sloping, emerged shelf. The animal community would have included mammals, birds, reptiles, and fish. Large animals probably included deer, bear, wolves, moose, caribou, bison, and musk ox. Elephant teeth that are often dredged from the sea floor of the continental shelf (Fig. 50) show that during the Laurentide glaciation, and perhaps as recently as 10,000 years ago, mastodons and mammoths roamed the region of the emerged continental shelf. As the Laurentide ice sheet retreated and the climate warmed, the plants and animals slowly migrated northward to occupy the Cape and Islands and the rest of New England. Man probably didn't live on the emerged continental shelf during full glacial times because it is likely that he had yet to complete his migration from Asia to eastern North America. However, there is evidence of early man in New England about 11,000 years ago; at that time, he undoubtedly lived on the continental shelf, as sea level was still well below its present level. Most likely he found the climate amenable and food plentiful. He probably hunted mastodon and mammoth and may have contributed to the extinction of these animals, but he probably subsisted mostly on smaller game, fish, and birds. Although the Indians missed the glaciation of the Cape and Islands, they witnessed rapid and dramatic changes in the environment. As the climate moderated, the evergreen forests and tundra that first occupied the Cape and Islands died out. They were replaced by the hardwood forests

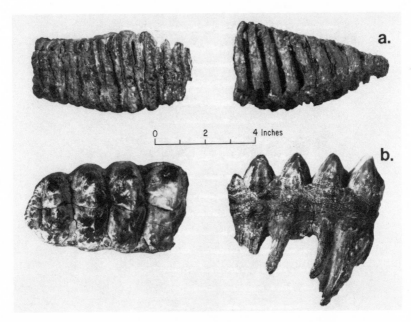

Figure 50. Mammoth (a) and mastodon (b) teeth dredged from the Gulf of Maine off Marblehead, Massachusetts. These teeth are similar to those dredged from the continental shelf south of the Islands and from Georges Bank, which indicates that mammoth and mastodon and other animals roamed the emerged continental margin during the Laurentide glaciation. Photographed by Frank C. Whitmore of the U.S. Geological Survey.

and other plants of a more temperate climate. Tundra animals such as caribou and musk ox moved north following the migration of their preferred food plants.

Pollen from cores in peat deposits suggests that when the Paleoindians first arrived in New England, they may have experienced a dramatic cooling as the climate swiftly returned to glacial cold between 11,000 and 10,000 years ago. However, the most dramatic change confronted by both the Paleoindians and early Indians was the submergence of the shelf as sea level rose.

On the gently sloping continental shelf, the sea's transgression was very rapid and may have been noticeable within a lifetime.

THE RISING SEA

Sea level rose as the great ice sheets melted and as the water trapped in the ice returned to the ocean basins. A great deal of ice melted during the early part of deglaciation and sea-level rise was very rapid, about 50 feet in 1,000 years (Fig. 51). After 10,000 years ago, most of the great ice sheets were greatly diminished, and the rate of sea-level rise slowed gradually until 6,000 years ago, when the ice sheets had disappeared from most areas of the globe. From that point until 2,000 years ago, sea-level rise was roughly 11 feet in 1,000 years. During the last 2,000 years, the general rate of sea-level rise has been about 3 feet every 1,000 years.

Tide gage records for the past 100 years suggest a rise in sea level of about 1 foot for the Cape and Islands. The cause of this increased rate of sea-level rise is not clear. Whether it will continue, slow down, or increase its rate, as proponents of the greenhouse theory claim, is unknown. Local tide-gage records indicate a rise in sea level, but this does not necessarily mean that there is an overall rise in the level of the world's oceans. Instead, the land may be undergoing local subsidence.

The landscape left behind by the glacier influenced the timing of marine submergence. The deep basins in the Gulf of Maine

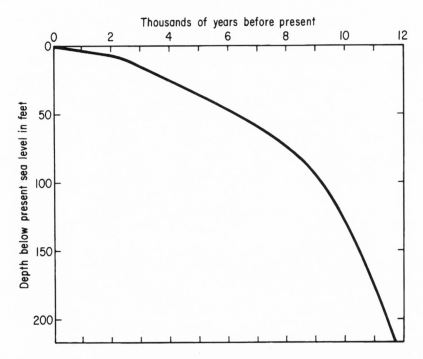

Figure 51. A sea-level rise curve for the last 12,000 years. As the Laurentide ice sheet and other ice sheets around the world retreated from their maximum advance, the sea began to rise as water trapped in the ice returned to the ocean basins. This curve is based on radiocarbon ages from shells, wood, and peat collected from cores taken on the continental shelf and from peat in the Great Marshes in Barnstable, Massachusetts. The initial rate of sea-level rise was rapid because the retreat of the ice sheets was rapid. As the ice sheets diminished, the rate of sea-level rise slowed; it slowed even more as the ice sheets in North America, Europe, and Asia disappeared altogether.

were flooded first and Cape Cod Bay next. Adjacent to the Gulf of Maine, marine submergence of the Cape and Islands may have begun shortly after ice retreat, and the final marine incursion into Cape Cod Bay began about 10,000 years before present. Vineyard Sound and Nantucket Sound were low lands and were flooded before the higher land on the emerged continental shelf south of the Islands. Marine waters entered the deeper parts of Vineyard Sound approximately 7,500 years ago and flooded Nantucket Sound about 6,000 years ago. About that time the Islands became separated from each other and from Cape Cod, and sea level continued to rise.

Thus, about 2,000 years ago, the Cape and Islands began to look something like they do today, even though the shoreline was probably a half a mile to several miles farther seaward. The great sea cliffs, with narrow strands of beach at their bases, the major barrier spits and islands, the lagoons and bays, and the salt marshes that characterize the Cape and Islands were well formed by that time.

MARINE DEPOSITS

When the sea transgressed the Cape and Islands, the glacial deposits became the raw material for the marine deposits. Waves and currents eroded, transported, and redeposited the drift to form the marine deposits beyond the shore and along the coast. These processes began as soon as the sea started to submerge the Cape and Islands and are still going on at the present time as they reshape the landscape and sea floor (Fig. 52).

Waves and currents have the same ability to sort sediments as running water on land. Thus, large waves and fast currents transport the coarser and heavier materials, and smaller waves and slow currents carry finer and lighter materials. Rock frag-

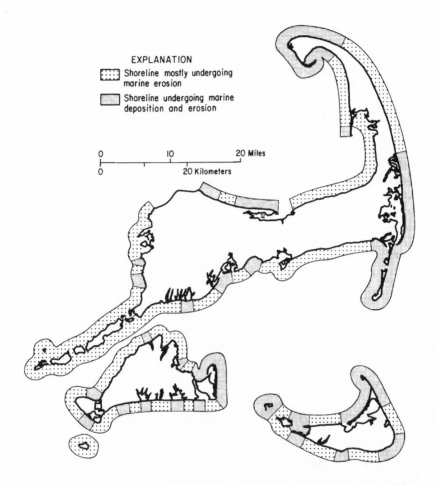

EXPLANATION

Shoreline mostly undergoing marine erosion

Shoreline undergoing marine deposition and erosion

Figure 52.   The coarse-stippled pattern shows shorelines around the Cape and Islands where the sea is eroding upland and forming sea cliffs or marine scarps. The fine-stippled pattern shows shorelines where sand is being deposited by waves and longshore currents. These depositional shorelines consist of spits, bay mouth bars, and barrier beaches. Although in places shorelines of marine deposition may be migrating seaward, in the long run, these shorelines will also retreat to keep place with the eroding headlands, thereby maintaining a straight coastline.

ments too large to be moved by the sea will be left behind. As the movement of the water slows, the transported material is deposited according to its size and weight. The coarsest or heaviest will be deposited first, and the finest or lightest, last.

The composition of the marine deposits is also controlled in part by the nature of the glacial sediments being eroded by the waves. Thus, if the raw material is either till or coarse ice-contact glacial drift, then the marine deposits will be composed of sand and gravel with numerous boulders; if the source is outwash, it will produce a sandy marine deposit with some pebbles and cobbles, but few boulders.

As the shoreline migrated landward across the continental shelf, it left behind an erosion surface that was cut by the waves. This surface, called the transgressive unconformity (Fig. 53), is forming today along the present shoreline, but it is as old as late Wisconsinan in deeper water offshore.

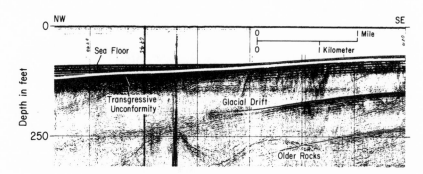

Figure 53.   As the sea rose to submerge the Cape and Islands, the waves eroded the glacial drift to form the transgressive unconformity, shown by the white line in this seismic-reflection profile, from Cape Cod Bay. The unconformity is time transgressive and is forming now along the shore as waves erode the land. In places, the unconformity forms the sea floor as shown on the right; in other places, it is covered by more recent marine deposits as shown on the left.

The marine deposits laid down beyond the shore rest upon the transgressive unconformity and consist of sand, gravel, boulders, and mud (composed mostly of silt and clay). The sand deposits usually include minor amounts of gravel and form a continuous blanket in many places, but locally the sand is piled high by the waves and currents to form shoals, such as Nantucket Shoals (Fig. 54). Mud may be deposited in deeper water where waves and currents do not reach the bottom. Concentrations of rocks on the sea floor, generally called ledges, are composed of boulders and gravel that were too heavy to be carried away by the waves and currents during the transgression.

Marine deposits that occupy the intertidal areas and bays and lagoons are generally sandy. However, they can be highly variable, and in some places they are composed of gravel or mud. The term "mud" means a mixture of silt and clay. In high-energy environments such as ocean beaches, the marine deposits tend to be composed of medium to very coarse sand with scattered, well-rounded, disk-shaped pebbles and cobbles. Deposits in lower energy, sea-floor environments such as shallow bays and the intertidal zone around bays and lagoons are commonly composed of finer sand and mud. Locally they may be very coarse, especially along shores bounded on the landward side by glacial deposits or where weak waves and currents transport only finer sand and leave behind a lag of pebbles and cobbles. Low-energy environments, where wave energy doesn't reach the sea floor and where tidal currents are weak or absent, are generally characterized by mud that contains abundant organic matter.

Tidal currents carry sand, silt, clay, and organic remains into the bays and lagoons either along the bottom or within the water column. As the tidal flow slackens, the sediments settle to the sea floor. Where tidal currents are strong, the bottom

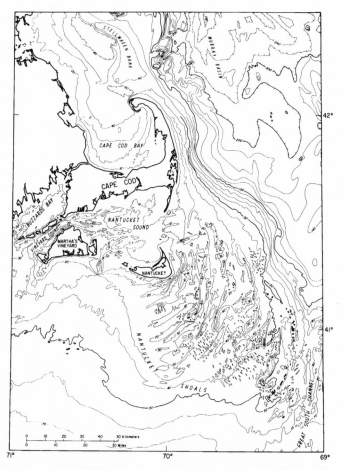

Figure 54.   Bathymetric map of the Cape and Islands region. Contours in meters below sea level (one meter is roughly equal to 3.28 feet). Nantucket Shoals, between Nantucket and Great South Channel, consists of numerous sand ridges formed by waves and tidal currents. The sand ridges overlie the transgressive unconformity. The shoals are a hazard to ships rounding Nantucket. Their threat caused the Pilgrims to turn back from their intended destination of Virginia and led to the founding of Plymouth in 1620. Nantucket and Vineyard Sounds are characterized by numerous sand ridges that are formed mostly by tidal currents. From a 1987 1:500,000 scale map by Elazar Uchupi of the Woods Hole Oceanographic Institution.

sediments are reworked by each tide and consist mostly of well-sorted sand. Broad areas of sand in the intertidal zone are called sand flats. Perhaps the best example is the broad, gently sloping sea floor in the southeast corner of Cape Cod Bay, where the shoreline migrates back and forth a mile or more twice a day as the tides ebb and flow. During the tidal transgressions, and during storms when the waves are large, the sand-flat deposits are reworked and transported, forming sand waves and ripple marks (Fig. 55). Where tidal currents run slowly, mud and finely divided organic matter are deposited on the sea floor.

Beach deposits are formed of material that was initially eroded from the glacial drift and transported offshore and along the shore. The waves return sand grains, pebbles, and cobbles to the beach, where they are deposited as the waves slow their rush up the beach.

Because the wind direction is often at an angle to the shore, the wind-driven waves usually rush up the beach obliquely and, under the force of gravity, return to the sea directly down the beach slope. In this way, the sand grains and gravel are transported in a saw-tooth path along the beach (Fig. 56). This motion of water along the beach is called a longshore current, and the resulting movement of beach materials is called longshore drift. As the sand grains are transported along the beach, they are sorted by size and density (weight per unit volume).

Grains of quartz are by far the most abundant constituent of beach sand. They have a low density and are transported most easily. Other sand grains are composed of minerals that have a high density compared to quartz. These are called heavy minerals. On the beach, waves and wind winnow the lighter quartz grains, and the heavy minerals are left behind. The heavy minerals, for example garnet and magnetite, are darker than

Figure 55. The sea floor off Brewster on Cape Cod shoals gently so that at low tide the sand flats are exposed for more than a mile offshore. The sand ridges are formed as wave action and tidal currents transport the sand to and fro. On the gently sloping sea floor, the rising tide advances swiftly, at times trapping swimmers and shellfishermen who unexpectedly find themselves cut off from the shore by deep water in the troughs between the sand ridges. Photograph by Dann S. Blackwood of the U.S. Geological Survey.

quartz sand, and concentrations of heavy minerals can be seen as dark patches on the beach. The difference in weight between quartz sand and heavy-mineral sand becomes readily apparent if a handful of each is picked up.

Finer-grained sediments, fine sand and mud, are carried into the intertidal zone where they are trapped by marsh vegeta-

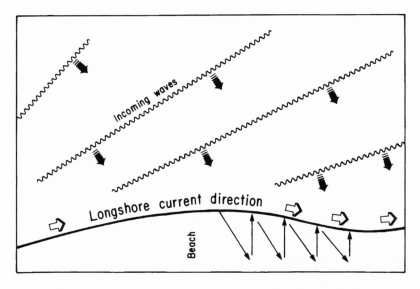

Figure 56.   Waves approaching the coast obliquely cause longshore currents and longshore drift. Sand is carried diagonally up the beach by the rush of the waves and directly down the beach by the back-wash. Sand carried into the water is transported along the sea floor by the longshore current. In these ways, longshore drift moves the sand along the beach below sea cliffs and along the barrier spits to their terminus, where they prograde into deep water.

tion. Coarser sand and even gravel can be included in the marsh deposits when storm waves breach a barrier beach or spit on the ocean side of the marsh, or they can also be eroded from glacial deposits around the landward side of a marsh. In some places, the sand layers within the marsh deposits represent a long record of major northeast storms and hurricanes. Some coarse material is carried into the marsh by sea ice that forms in shallow water and traps debris in its base. Blocks of sea ice float into the marsh on the tides and leave behind the sand and gravel when they melt. All of these sediments mix with abundant dead and decaying marsh grass to form the marsh deposits (Fig. 57).

Figure 57.   Salt-marsh peat deposits are a mixture of decaying or-
ganic matter (mostly marsh grasses), sand, and mud. The sand and
mud was carried into the marsh by tidal currents and deposited
among the grass plants. Soil horizons in salt marsh peat are difficult
to see because they lack contrast in the generally black to dark-
brown soil and parent material. The soil horizons are weakly formed
because the peat is being continuously deposited to keep pace with
the rising sea level. Photograph by Peter Venerman of the University
of Massachusetts, Amherst.

Salt-marsh plants grow at or near sea level. As sea level rises, the plants are drowned and become covered by sediment and by new marsh plants. In this way, the marsh keeps pace with rising sea level and at the same time migrates landward.

Large boulders that can be seen in many marshes may seem incongruous, and they are not part of the marsh deposit. They are glacially deposited boulders that are in the process of being buried by the rising marsh surface. Similarly, tree stumps can be found in the marsh in some places. At one time these trees were growing upon an upland surface that was later submerged and buried by the accumulating marsh deposits.

WIND DEPOSITS

Strong onshore winds pick up sand grains from the beach or cliff face and carry them inland, where they are deposited to form dunes. The grains are moved downwind in a series of hops, a process that is called saltation. Many sand grains are in the air at the same time, and they form a moving carpet above the beach or dune surface. During strong storm winds, the grains move at high speed, as anyone who has been on the beach at such times and has been stung by flying sand grains can attest. As the wind passes inland, its velocity slows and the sand grains drop to the dune surface. Some grains are carried over the crest of the dune and are deposited on the downwind face. When enough sand has been deposited on the steep downwind face of the dune, called the slip face, it slides to the base of the dune. The processes of saltation and sliding cause the dune deposits to migrate in the direction of the wind.

The winds cannot carry pebbles and cobbles and therefore the dune deposits are composed only of sand (Fig. 58). Pebbles and cobbles associated with dune deposits owe their existence to other factors. In many places they occur as a lag deposit that marks the base of the dune deposits and the top of the underlying beach deposits. The lag deposits form when the wind

Figure 58. Sand pit in the Mount Ararat dune, Provincetown and Truro. Sand layers within the dune appear horizontal, but they are slip face deposits that dip steeply toward the viewer. The beds appear horizontal because the pit face is normal to the dip of the beds. Sand from this pit was used for construction, foundry molds, and making glass. About 60,000 tons of sand were removed each year during the approximately 50-year operation of the pit. The pit face is about 25 feet high. Photograph by Stephen P. Leatherman of the University of Maryland.

removes sand from the beach and leaves beach cobbles and pebbles behind.

Other coarse fragments were carried into the dunes by man. A concentration of cobbles and small boulders may mark a hearth constructed by Indians in the distant past. Sometimes fragments of pottery, bones, and other artifacts are found among the hearth stones or scattered nearby. Many of these hearths are associated with dark layers within the dune sand that are fossil soils, which were covered by drifting sand (Fig. 59). When these paleosoils are exhumed, as overlying sand deposits are eroded by the wind, fossil stumps and roots are commonly found on the surface. These paleosoils represent times when mature forests covered the dunes.

Figure 59. Fossil soil exposed in the Provincelands dunes. The soil represents a time when the dune surface was stable and well vegetated by shrubs and trees. In some places, Indian hearths and artifacts are associated with fossil soil horizons. The forest was killed when migrating dunes buried it and the soil. Photograph by Stephen P. Leatherman of the University of Maryland.

COASTAL LANDFORMS

Marine coastal erosion and deposition have added distinctive landforms to the glacial Cape and Islands. These landforms include sea cliffs, beaches and barrier islands, marshes, coastal ponds, and sand dunes, which along with the glacially generated landforms make up the landscape.

Marine Scarps

One of the most dramatic landforms is the marine scarp or sea cliff (Figs. 16, 60). The most spectacular marine scarps, for example, Gay Head Cliff (Fig. 14), Sankaty Head Cliff, and the cliff along the ocean shore of lower Cape Cod from Eastham to Truro, rise abruptly from the sea to reach heights of 100 to 150 feet. The great cliff along the ocean side of outer Cape Cod is perhaps the most impressive of the marine scarps to be seen on the Cape and Islands. The cliff is more than 15 miles long and would form an impenetrable barrier to movement inland if it weren't for the "hollows" that interrupt the cliff at irregular intervals. The hollows are the upstream parts of the dry valleys that cross the outwash plains. Today they provide access to the beach for swimmers, windsurfers, and surf fishermen. In the past, when sailing vessels became stranded along the ocean side of the outer Cape—not an uncommon occurrence, for ships were wrecked every two or three weeks on average—the hollows provided a way off the beach to stranded sailors and

Figure 60.   Sea cliff or marine scarp at Sankaty Head on Nantucket. Marine scarps provide unequivocal evidence of coastal erosion and are the main source of sediment for the marine deposits and most of the sand dunes. For geologists, marine scarps expose layers that provide much of the evidence of the geologic history of the Cape and Islands. For example, drift of probable Illinoian age, the marine Sankaty Sand of Sangamonian age, and drift of late Wisconsinan age are exposed in the Sankaty Head Cliff; Gay Head Cliff exposes marine and terrestrial strata that provide most of the evidence for events that occurred during Cretaceous and Tertiary time.

access to the beach for lifesavers and wreckers. In many other places, the marine scarps are lower.

All marine scarps provide clear evidence that the glacial Cape and Islands are being worn away by the sea. They form wherever the glacial deposits face the open sea and are unprotected from wave attack. The marine scarps show where the coastline has been smoothed by the erosion of headlands, which must have characterized the coast as submergence began. Erosion of the marine scarps is episodic, occurring during northeast storms and hurricanes. Even then the marine scarps may undergo erosion in some places and not in others.

Where the marine scarp is underlain by incohesive sandy deposits, storm waves attack the base of the scarp and remove a wedge of material, leaving a near vertical face. Later, during quieter times, the sand collapses until the scarp develops a slope of about 30 degrees or less. Thirty degrees is the maximum slope that can be maintained in incohesive sand and is called the angle of repose.

Marine scarps in more cohesive deposits, for example, deposits that contain clay, erode differently. Under quiet conditions, the clay scarp will be much steeper and in places, even after the erosion has temporarily ceased, will remain nearly vertical. Usually over time the scarp will fail by slumping or, if water saturated, by earth flow. These types of slope failure occur at Gay Head Cliff, where there are clay deposits of Cretaceous age, and at Highland Light Cliff, where a thick clay layer occurs between layers of sand. Evidence of slumping can be seen at the tops of these cliffs, where material moving downslope makes up a series of steplike blocks. The slumped material accumulates at the base of the slope and partly stabilizes the scarp. Storm waves remove this buttress, and the process of slumping and land sliding begin again.

Although erosion and landsliding result in a loss of land to the sea, they are necessary to maintain the marine scarp. Years ago, concern about the loss of land at the Gay Head Cliff led to an investigation on how to control the erosion. The conclusion was that any effort to stop the erosion would eventually lead to the loss of the cliff as a natural wonder when the multicolored strata became covered by slump material. When large storm waves attack, the cliffs erode rapidly. At these times, the cliff face retreat can be on the order of tens of feet in each storm. For example, the bluff at Nobska Point in Woods Hole was eroded back as much as 50 feet during the September 1938 hurricane. At other times, the marine scarps may be protected from erosion by the accretion of beach sand along the shore, by the formation of barrier beaches, or by the development of offshore bars that dissipate wave energy well offshore.

The sea cliffs may be protected from erosion for many years, during which time they will become well vegetated, even forested. Settlements may be built at the top, on the face, and at the base of well-protected fossil sea cliffs. The village of Nantucket is built on a sea cliff that has been protected for hundreds of years from the open sea by Great Point and Coatue spit. On the other hand, Siasconset village is built above a marine scarp that faces the open Atlantic Ocean. This scarp is presently protected by an accreting beach, but this has not always been so. Siasconset village was founded in 1695 and includes some of the oldest houses on Nantucket. However, this does not indicate a persistently stable shore. During a number of times since the founding of the village, the cliff face was unprotected from wave attack and eroding. At these times, houses were moved back away from the sea cliff to safety.

Thus, in the long run, the protection offered by accreting beaches, barrier spits, and offshore bars is temporary. This fact is shown by the renewed erosion of the vegetated and stabilized marine scarp along the mainland shore at Chatham after the recent breakthrough at North Beach, the barrier spit between Chatham and the open ocean, and by the renewed erosion at Sankaty Head, Nantucket.

Beaches and Barriers

Beaches form as narrow strands below sea cliffs and as barriers that separate lagoons and embayments from the open ocean. During fair weather with gentle winds and quiet seas, waves and longshore currents bring sand to the shore to build a beach with a gently sloping profile that looks much the same along its entire length. This is called the summer beach, and, to most summer visitors, the beach looks much the same from day to day. However, strong winds and high seas occur commonly at other times of the year, during occasional hurricanes and the more abundant northeast storms. During these events, large storm waves attack the beach and return great amounts of sand to the sea. At these times, the beach changes rapidly. The beach profile steepens and has vertical scarps up to a few feet high. Since these changes occur mostly in the winter time, this is called the winter beach.

The beach can be divided into parts (Fig. 61). Adjacent to the beach on the seaward side is the shoreface, a seaward-sloping surface with longshore troughs and longshore bars that remain submerged except during the very lowest tides. The foreshore is the part of the beach that is normally covered and uncovered by the tides, and the backshore is the part of the beach that is normally above high-tide level.

Coastal barriers (Fig. 62) are long and narrow landforms that are bordered on the seaward side by the open ocean and on the landward side by bays and lagoons. From place to place, the coastal barriers are interrupted by inlets, and most are re-curved or hooked at their distal end. Salt marshes, tidal flats, and flood and ebb tide deltas make up the tidal and subtidal parts of a coastal barrier. When a coastal barrier is detached from the mainland, it is considered to be a barrier island and when attached, a barrier spit. These conditions tend to be

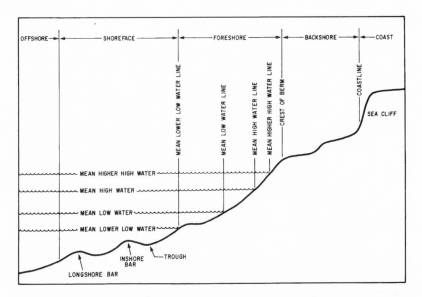

Figure 61. *A depiction of a beach profile from offshore to the marine scarp showing the parts of a beach and the relation between tide levels and the morphologic features found in a beach. The shoreface represents the part of the beach that is affected by waves and currents that build bars parallel to the shore. It remains submerged except during extreme low tides. The foreshore is the part of the beach normally covered and uncovered by the tides. The foreshore surface is generally flat because of the rush and backwash of the breaking waves. The backshore is generally above the reach of the tides and waves. It protects the sea cliff from wave attack except during major storms when the tide is unusually high and the waves larger than normal. At these times, rapid erosion of the cliff face may occur.*

temporary as some barriers attach and detach over and over again; thus, barrier islands and barrier spits can simply be called barriers. The major barriers on the Cape and Islands include the Provincetown Spit, Nauset Beach, and Monomoy Island on lower Cape Cod; Sandy Neck on the north shore of the upper Cape; Cape Poge on Martha's Vineyard; and Great Point and Coatue on Nantucket.

Figure 62.   Embayments on the Cape and Islands are protected by barrier spits and barrier islands. This southward-looking aerial photo, taken over Nauset Inlet, shows the barrier spits that protect Nauset Harbor and Nauset marsh. Further to the south, a barrier protects Pleasant Bay. The break in the barrier off Chatham that formed in 1987 can be seen in the upper center of the photograph. Photographed by Dann S. Blackwood of the U.S. Geological Survey.

The subaerial part of the barriers consists of beach sand capped by sand dunes. Because the longshore currents and longshore drift continue on in the same direction as they leave the cliffed sections of the coast, the barriers tend to be straight and to close off the irregular upland coast from the open ocean. The distal ends of barriers generally recurve or hook. The hooking or recurving is caused by refraction of the waves around the end of the barrier, which, in turn, changes the direction of the longshore current and carries the longshore drifting sand around the end of the spit. In some places, the hook partly encloses a bay, for example, Provincetown Harbor and Cape

Poge Bay on Martha's Vineyard. Eventually these bays may be closed off from the sea to become a fresh- to brackish-water lake. As the barrier increases in length, a new recurved spit will form, and the older recurved spit will be left to become part of the main barrier. These fossil recurved spits mark the former positions of the barrier terminus and can be recognized as series of arcuate lines of dunes that may enclose a series of ponds (Fig. 63).

Barriers are dynamic features that are continually changing shape and position. Barriers migrate laterally as sand is carried by the longshore drift and deposited into deep water at the distal end. They also migrate both seaward and landward. The seaward migration called progradation occurs when sand is deposited along the foreshore and shoreface. Landward migration occurs when storms erode the foreshore, shoreface, and backshore and as storm waves overtop the dunes to carry beach and dune sand into the adjacent bay, lagoon, sand flat, or marsh. The primary means of landward migration is called overwash. Flood-tide deltas at inlets also contribute to the landward migration of the barriers as tides carry sand into the sheltered water where it is deposited and becomes the base for the beach and dune. Under quieter conditions, the surface of the barrier is modified by winds. Beach and dune sand are picked up and carried forward to form dunes on the new surfaces formed during the storm. By these processes, the barriers migrate landward. Landward migration is most pronounced and rapid where sea level is rising, a condition that presently occurs on the Cape and Islands and that may increase in the future, if a sea-level rise caused by global warming is a reality.

Major storms may breach the barrier. A breach formed during the famous 1978 northeaster broke Monomoy Island into two parts. More recently, in 1987, people in Chatham awoke the

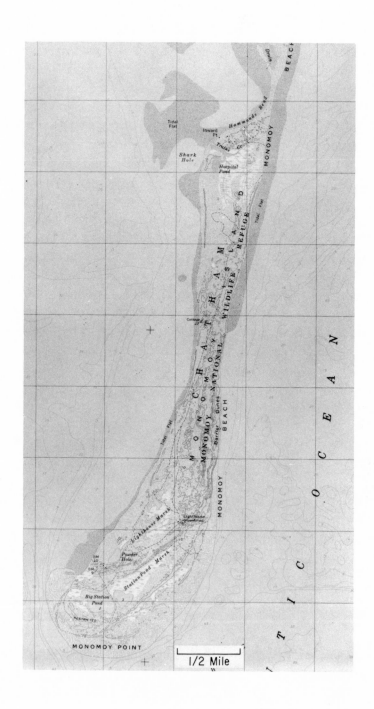

1/2 Mile

120

morning after a northeaster to find a breach in Nauset Beach (Fig. 64) and their property no longer protected from ocean waves. Many storm-generated breaches heal quickly, but the one in North Beach at Chatham has widened greatly and is migrating southward, exposing more and more of the upland to wave attack.

Why doesn't every major storm cause breaches in the barrier? Storm waves alone probably cannot cause a breach. Other factors are involved, the most important of which may be a higher water level in the lagoon than along the ocean side of the barrier. The different water levels may be caused by a difference in the time of high or low tide augmented by higher or lower tides across the barrier caused by storm winds. On Martha's Vineyard, geologist Gordon Odgen observed the formation of a breach in the barrier that separates Katama Bay from the ocean to the south. The barrier was being attacked by storm waves generated by Hurricane Edna, which occurred in 1954. The waves swept across the crest of the barrier and leveled sand dunes as high as 20 feet. At that time the tide was ebbing along the ocean side of the barrier, and Katama Bay stood about 5 feet above the ocean level. At one point, he stepped across a small scarp in the beach and sank in sand up to his knees, the result of water flowing from the bay to the ocean through the barrier and turning the beach sand quick.

Figure 63. (opposite)   Part of the topographic map of the Monomoy 7½-minute quadrangle (original scale 1:25,000) showing a recurved spit. This example is at the southern tip of Monomoy Island at the east end of Nantucket Sound. The recurving is caused by the wave refraction at the tip of the spit. Big Station Pond and Powder Hole were once harbors protected by former recurved spits now represented by lines of dunes. The southern end of Monomoy is prograding eastward as well as southward and the lighthouse, once right at the shore, is now more than a quarter of a mile inland.

122

The beach sand was being liquefied by discharging water fed by the high water level of the bay. He quickly retreated to firm ground and observed the storm waves remove more and more of the quicksand until the barrier was breached. Water began to flow from the bay to the ocean and within 2 hours the breach was almost 1,000 feet across.

Salt Marshes

Salt marshes abound on the Cape and Islands. They occur where the wave energy is low; for example, behind the barriers, along the upland shore of bays, and along the sides of tidal creeks. Marshes have a generally flat surface that is cut by complexly meandering tidal creeks and depressions called pond holes or pannes (Fig. 65). Numerous ditches have been cut in the marsh surface in an attempt to control salt-marsh mosquitoes by preventing standing stagnant water on the marsh. The ditches were mostly dug during the 1930's and their effect on the overall ecology of the marsh is poorly understood. Some studies suggest that ditching caused major changes in the plant and animal population of the marshes. An effort has been made to restore some marshes in order to produce a more favorable habitat for the mummichog minnow. It is hoped that the mummichog will eat the mosquito larvae and thereby control the mosquito population.

Figure 64. (opposite)   The Chatham break occurred on Friday, January 2, 1987. The upper photograph shows the break two days after it was formed. An off-road vehicle at the north tip provides scale. Photograph by Kelsey and Kennard. The lower photograph shows the break during the summer of 1991, photographed at near low tide. The break is now about a mile wide; at low water, a flood-tide delta closes the channel that was used before the break to enter and leave Chatham Harbor. The 1991 photograph by Dann S. Blackwood of the U.S. Geological Survey.

Figure 65. Aerial photograph of Sandy Neck Spit and the Great Marshes at Barnstable. Material to build Sandy Neck comes from the cliffed shoreline to the west. Longshore currents and longshore drift move the sand eastward. The spit protects the marsh and Barnstable Harbor from the open sea. Tidal creeks, salt pannes, and mosquito-control ditches interrupt the marsh surface. The spit and marsh began to form about 4,000 years ago as the rising sea began to drown the glacial Cape. The upper Cape, Nantucket Sound, and Monomoy Island can be seen in the background. Photograph by Dann S. Blackwood, U.S. Geological Survey.

Marshes are classified according to the frequency of flooding by high tides. The low marsh is flooded twice a day on every high tide, while the high marsh is flooded much less frequently, usually during spring tides when the tide is much higher than normal. Each has its own type of vegetation. Low marsh is characterized by a grass called *Spartina alterniflora*, also called cord grass. The high marsh is dominated by *Spartina patens*, also a cord grass, which is the source of salt-marsh hay. The hay was an important crop in the past, when agriculture, rather than tourism and real estate, was a major part of the economy of the Cape and Islands.

The oldest salt-marsh deposits on the Cape and Islands are between 4,000 and 6,000 years old, and they appear to correlate with the slowing of sea-level rise. Before that time, the submergence of the region may have been too rapid for extensive marsh deposits to form. The remains of older marsh plants and Foraminifera (small single-celled animals with a hard shell) that live only in the marsh environment can be sampled by coring. Their organic matter can be dated by radiocarbon analysis and can provide a record of sea-level rise over thousands of years (Fig. 66).

The salt marshes on the Cape and Islands form in sheltered bays and estuaries where accumulations of sand and mud shoaled the sea floor to intertidal depths. As the area of shoal water increases, the marsh advances toward the deeper parts of the embayment. With a rising sea level, the marsh also advances over the upland. Stages in the development of the Sandy Neck barrier and the Great Marshes in Sandwich and Barnstable on Cape Cod have been determined from radiocarbon ages obtained from the marsh sediments. About 3,000 years ago, the barrier was a little more than 1 mile long. It protected a small lagoon and a few patches of marsh. As the barrier lengthened and the lagoon shallowed because of sedi-

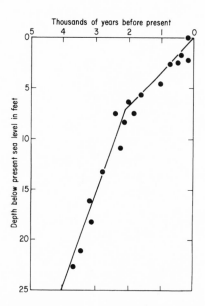

Figure 66.   Sea-level rise curve (constructed by Alfred C. Redfield of the Woods Hole Oceanographic Institution) for the Great Marshes in Barnstable on Cape Cod. Black circles represent the age and depth of radiocarbon-dated samples of salt-marsh peat. Each peat sample comes from the base of the marsh deposits and represents a former sea level. The oldest sample was obtained from the deepest marsh deposit, and its age represents the time when the marsh first formed. Because the marsh requires sheltered water, the age of the oldest sample also approximates the time when Sandy Neck Spit began to form.

mentation and as sea level rose, the marsh increased greatly in size. Today the barrier is more than 6 miles long and protects a large lagoon and a marsh of several square miles (Fig. 67).

Coastal Ponds

Coastal ponds occur in many places along the shore of the Cape and Islands. In general they are characterized by a sheltered body of salt, brackish, or fresh water closed off from the open sea

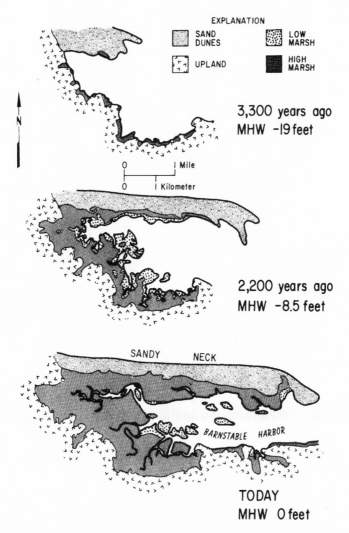

Figure 67. Development of Sandy Neck spit and the Great Marshes of Barnstable. The stages were identified by Alfred C. Redfield of the Woods Hole Oceanographic Institution. Between 3,300 years ago and today, the spit has grown to a length of over 5 miles. As the spit increased in length, peat deposits formed in the sheltered lagoon. Continued growth of the spit, the accumulation of sand transported by the tides, and the rise in sea level will cause the lagoon to shoal and the extent of high marsh to increase. (MHW = mean high water.)

by a beach called a baymouth bar. A few have a natural opening to the sea, but many are connected to the sea by a manmade channel. Coastal ponds occupy kettles or the lower part of outwash plain valleys. Kettles form some salt ponds and bays. The kettle holes have been partly submerged by the postglacial rise in sea level and breached by the sea, for example, Waquoit Bay on inner Cape Cod, Salt Pond in Eastham, and Wellfleet Harbor. The coastal ponds that occupy the outwash plain valleys are usually closed off from the open ocean by baymouth bars. The coastal ponds may be broad, like the ones on the south shore of Martha's Vineyard. Some are long and narrow, like the coastal ponds along the southshore of upper Cape Cod. Other bays owe their existence to minor sublobes of the Laurentide ice that acted like large ice blocks. The sublobes excluded glacial drift and, as they retreated, caused any overlying sediment to collapse to form topographic lows. Pleasant Bay and Little Pleasant Bay in Chatham and Orleans and Nauset Harbor in Orleans and Eastham mark sublobes of the South Channel lobe. Coastal ponds that occupy kettles owe their existence to sea-level rise and to coastal erosion that caused the shoreline to retreat until it intersected the kettle. Cores taken in the bottom of these bays show saltwater deposits overlying brackish-water deposits that in turn overlie freshwater deposits. The change from fresh water environment to salt or brackish-water environment depends upon the time when a threshold was overtopped by the rising sea or when the retreating shoreline breached the kettle. The earliest change from freshwater to a coastal pond probably was about 4,000 years ago or about the same time that the earliest of the salt marshes formed. Coastal ponds that occupy the lower parts of outwash plain valleys may owe their existence solely to the transgression of the sea because they are open to the sea and were graded to a lower sea level.

Other topographic lows now partly drowned by the sea may reflect changes in elevation in the surface that underlies the glacial drift. In many places, offshore seismic profiles show that topographic lows in the glacial drift surface correspond to lows in the surface that underlies the drift. The topographic sags between Gay Head and Chilmark that contain Menemsha Pond and Squibnocket Pond, the one that contains Vineyard Haven Harbor, and the one that contains Edgartown and Katama Bay all developed over lows in the preglacial surface atop the coastal plain deposits. The passages through the Elizabeth Islands, called holes, for example, Quicks Hole and Woods Hole, may also have developed over lows in the surface beneath the drift.

Many coastal ponds have been managed since colonial times, when the settlers discovered that temporarily breaching the baymouth bar increased the salinity of the pond and provided a good habitat for shellfish, particularly oysters, and finfish. Additional benefits to opening a pond to the sea included a reduction in the mosquito population and a reduction of aquatic vegetation that would lower the oxygen content of the water if unchecked. For some time, modern wetland regulations prevented the temporary opening of coastal ponds to the sea. However, a severe degradation in the quality of the water in a number of coastal ponds may lead to a decision to reopen them to allow an influx of saltwater.

Sand Dunes

Coastal sand dunes have formed in conjunction with the growth of the barrier spits and barrier islands. They have also formed atop sea cliffs cut in the glacial drift. The most extensive dune fields are found on Provincetown Spit, on Sandy Neck Spit in Barnstable, and on Monomoy Island south of the elbow of Cape Cod. Smaller dune fields occur on Cotue and Great Point Spits on Nantucket and along the spit that makes up the east side of Martha's Vineyard. In most places, the sand dunes are only a few feet to a few tens of feet high, and the relief of the dune field is low. However, on Sandy Neck Spit, the dunes reach altitudes of 70 feet, and, on Provincetown Spit, some dunes are more than 100 feet high.

To someone walking through the coastal dunes, they may appear as random hills of sand separated by low areas called blowouts. However, when viewed from above, from over-looks, or on aerial photographs and topographic maps (Fig. 68), many of the dunes can be seen to form linear or parabolic dune ridges. The linear ridge nearest the beach is called a foredune and is parallel to the shore. In a dune field, it is the youngest dune ridge and is being formed as onshore winds carry sand inland. Older inland dune ridges that are either

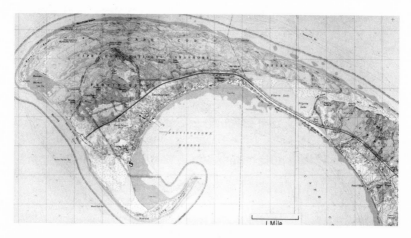

Figure 68.   Topographic map of the Provincetown 7½-minute quad-rangle and part of the North Truro 7½-minute quadrangle. Original scale 1:25,000. The topographic map shows the parabolic dunes north of Pilgrim Lake, the foredune along the shore, and the linear dune ridges north of Provincetown. The linear ridges may mark former shorelines of the spit, which is prograding into water depths of up to 200 feet. Mount Ararat, the highest dune on Provincetown Spit, has an elevation of about 100 feet.

parallel or subparallel to the present shore may represent previous foredunes formed along older shorelines left behind by the progradation of the spit or barrier. Many of these older inland dunes are active and make up the largest and highest dunes. The best-developed parabolic dunes are on the landward end of Provincetown Spit (Fig. 69), and they also occur on

Figure 69. Aerial view from the west of the parabolic dunes on Provincetown Spit. The prevailing westerly winds have blown out the centers of the dunes, and so the dunes open to the west. The southern limbs of the parabolic dunes are advancing on Pilgrim Lake. The lake was once called East Harbor, which was an embayment open to the sea. Shoaling by sand blown into the harbor and the possibility that this sand would eventually shoal Provincetown Harbor caused the inhabitants to close the harbor entrance in 1869. This photo taken in 1991 shows beach grass planted in an effort to stabilize the dunes. Experience has shown that the beach grass must be replaced by trees and shrubs before the dunes will become stable and eventually be covered by soil and forests. Photograph by Dann S. Blackwood of the U.S. Geological Survey.

Sandy Neck Spit. Parabolic dunes are U-shaped and the opening faces the strongest prevailing winds, which, at Provincetown and Sandy Neck Spits, are westerlies. The highest part of the dune is at the bottom of the U, and the limbs gradually decrease in height. The low in the center of the parabolic dune is a blowout carved from an existing dune by the wind. In some places, a lag of beach pebbles and cobbles on the depression floor shows that the dune sand has been completely removed to expose the beach deposits that make up the foundation for the sand dunes.

Developing dunes have no soil and little or no vegetative cover because erosion of sand from the windward side and deposition of sand on the leeward side prevent a soil from forming and most plants from growing. By the processes of erosion and deposition, the dunes migrate before the wind and advance into forests, salt marshes, lakes and ponds. In Truro on lower Cape Cod, the dunes are filling Pilgrim Lake. In Provincetown, a sand dune is migrating across Route 6, and the sand must be removed periodically to keep the road open. In some places, the dunes are stabilized by soil and vegetation and are inactive. It is said that the older dunes of the Provincetown Spit were well covered by soil, trees, and grass when the earliest settlers arrived. However, this protective cover was quickly destroyed as the trees were cut down to provide fuel and lumber and the grass was overgrazed. The newly exposed sand was attacked by the wind, and the dunes began to migrate. Beginning in 1662 numerous local and state laws were passed to prevent further destruction of the soil, trees, and grass. These laws were largely ignored or were unenforced, and the destruction of the forest cover continued. Late in the 18th century, the migrating dunes began to bury Provincetown village; only then were more serious efforts made to stabilize the dune surface by planting beach grass and trees. However, even today sand dunes are encroaching on the forest in some places (Fig. 70).

Figure 70.   Dune encroaching on forest on Provincetown Spit. Despite numerous laws going as far back as 1662 and many attempts to stabilize the dunes with vegetation, the dunes are still on the move. The recently planted beach grass (Fig. 69) may slow the advance, but reforestation must occur before the dunes are fully stabilized. Photograph provided by the Cape Cod National Seashore.

Buried soils (Fig. 59) and "fossil" trees within the dunes represent episodes of renewed dune migration caused by deforestation due to natural causes such as forest fires or to human causes such as cutting and overgrazing. From place to place on Cape Cod, small fields of forested dunes occur far from the shore. The dunes overlie the soil developed on the drift and have an immature soil on their surface. The age of these inland dunes is not known, but they probably formed when land was cleared and plowed to support the Cape's agrarian economy.

SOILS

Soils on the glacial drift, the late-glacial windblown sand and silt and dune sand, are called mineral soils. Mineral soils consist mostly of mineral grains along with variable amounts of organic matter. Soils begin to form on the glacial and windblown deposits soon after deposition ceases and the surface becomes stabilized. The upper part of the deposits begins to weather—to disintegrate because of physical and chemical processes caused by exposure to air and water. Plants slowly begin to colonize the glacial landscape, and plant litter starts to accumulate on the ground surface. Water from rain and melting snow percolate downward through the plant litter and the glacial deposits. The water carries finely divided organic matter, minerals, and chemicals leached from the plant litter and mineral matter, silt, and clay from the upper part of the glacial deposits and redeposits them a short distance below the surface. Uprooting of trees and burrowing by animals, including earthworms and ants, are important soil-mixing processes, as are cycles of freeze and thaw.

These soil-forming processes are controlled by several factors, including time, climate, plants and animals, topographic slope, and the nature of the deposit—called the parent material—on which the soil develops. Time is a factor because soil formation is a slow, continuing process. Thus, the mineral soil on the glacial drift of the Cape and Islands began to form about 18,000 years ago and is much better developed than the mineral soils that have formed on the much younger dune deposits. Climate factors involve the amount and type of precipitation, the number of freeze and thaw cycles, and temperature that controls the rate of chemical change. Plants and animals influence soil texture, permeability, and soil mixing. Bacteria and fungi affect soil chemistry and promote decay in the plant litter. Slopes control the amount of percolation and runoff and soil moisture. Soils on steep slopes tend to be well drained, whereas those on gently sloping surfaces may be poorly drained, especially where they are near the water table. The parent material is the major factor in the mineralogy and soil texture. For example, soils developed on till or ice-contact deposits are stony, and those formed on outwash are mostly sandy. Glacial drift and late-glacial eolian deposits are the most extensive parent materials for the mineral soils on the Cape and Islands. Parent material and slope are the most important factors in the formation of mineral soils on the drift and late-glacial eolian deposits because they are highly variable. The remaining factors—for example, climate—are pretty much the same over the entire Cape and Islands.

The mineral soils on the Cape and Islands are classified as podzols. A podzol forms in regions where the climate is sufficiently cold to restrict chemical and biological activity and where trees are the natural vegetative cover. The typical soil profile in a sandy podzol (Fig. 71) is divided into three soil horizons. The uppermost horizon, called the A horizon, consists of, from top to bottom, a layer of organic litter undergoing decay, a middle layer that consists of mixed humus and mineral grains, and a lower layer that is mostly mineral grains. The layers rich in organic matter are generally light to dark brown to black. The lower layer is easily recognizable because it is bone white; it is being leached by percolating water. The middle horizon (the B horizon) is very distinctive because of its reddish-brown color, caused by precipitation of iron-bearing minerals that were leached from the A horizon. The lowest (C) soil horizon consists of the parent material. Because the mineral

Figure 71.   Typical Cape and Islands upland podzol soil, showing, from top to bottom, the organic litter zone, the mixed organic and mineral-matter zone, and the leached (light-colored) zone. Together these zones make up the A horizon. Below the A horizon is the darker (orange-brown) B horizon, where iron-bearing minerals leached from the light-colored layer of the A horizon are precipitated. The C horizon, the parent material, does not show clearly in this photo-graph. A Cape Cod National Seashore photo.

soil develops progressively, over time the soil layers will grade into each other and become thicker with increasing age.

Much of the mineral soil on the Cape and Islands has been altered by man. In the past, many areas were agricultural, and the upper soil layers were mixed by plowing or were destroyed by overgraz-ing and lumbering. More recently the topsoil was mined and taken elsewhere, but this practice has been greatly curtailed. Today, many soils are modified during clearing and construction.

Swamp and marsh deposits make up the parent material for organic soils. Organic soils are not obviously zoned because they have a very high content of decaying vegetation that has a generally black color. They are also surfaces of active deposition, especially where salt marshes grow upward in response to sea-level rise; thus, most organic soils are immature.

The soils on the Cape and Islands have been mapped by the U.S. Department of Agriculture's Soil Conservation Service. The maps are much more complex than a geologic map, because each geologic unit is the parent material for numerous soil map units (Fig. 72). The soils maps for the Cape and Islands are published on a county basis (Appendix C).

Figure 72.   Sheet 8 (Orleans and vicinity) of the Barnstable County soils map. There are many more soils units than there are glacial drift units on the geologic map of the same area (Fig. 8), mainly because surface slope is a major factor in the soil classification. Beach, dune, and wetland soils mostly occupy the same positions on both the soils map and the geologic map. Sheet 8 was provided by Peter C. Fletcher of the U.S. Department of Agriculture, Soil Conservation Service.

FRESHWATER

The Cape and Islands are a water dowser's paradise because freshwater lies at variable depths everywhere beneath the land surface. Predicting its presence is a sure thing. The lakes and ponds that abound on Cape Cod are an indication of abundant freshwater beneath the surface. Contrary to local folklore, the water on the Cape and Islands is not an underground river that flows from the White Mountains of New Hampshire. The only source is local precipitation, and that, in turn, is dependent on the hydrologic cycle. The hydrologic cycle is the continuous circulation of water among land, sea, and atmosphere (Fig. 73). About half of the precipitation is returned directly to the atmosphere by evaporation and by transpiration from plants. The remaining half enters the ground and eventually enters the ocean. This subsurface water, called ground water, fills all the voids between the mineral grains in the unconsolidated deposits. The deposits that are saturated with freshwater are called aquifers.

The Cape and Islands are hydrologically simple in that they are composed mostly of sand and gravel, and they are completely or almost completely surrounded by the sea. In a very generalized way, the aquifer is made up of sand and gravel. In cross section, it is a broad, doubly convex lens-shaped body (Fig. 74). The top of the aquifer is marked by the contact between saturated and unsaturated ground, called the water table. The bottom of the aquifer is marked by the contact between fresh ground water and saltwater or the bedrock surface. The water

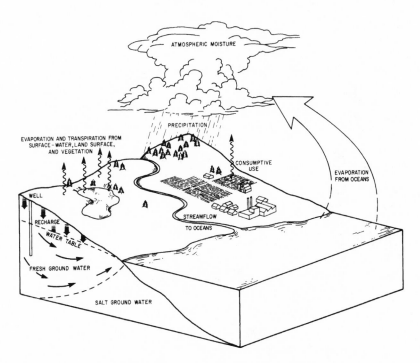

Figure 73. The hydrologic cycle. Freshwater on the Cape and Islands depends solely on the precipitation that falls locally. There is no underground river bringing freshwater to Cape Cod. Water that falls to the ground as rain or snow was mostly evaporated from oceans and lakes and enters the atmosphere as water vapor. It condenses to form clouds and falls to the earth as rain or snow. About half of the precipitation that falls is returned directly to the atmosphere through evaporation and transpiration from plants. The remainder percolates downward to replenish the aquifer. Ground water is then used by the population or returns slowly to the sea.

table surface is convex upward with its highest point beneath the highest ground surface. Similarly the lowest point on the saltwater-freshwater contact occurs below the highest point on the land surface. The horizontal limit of the aquifer is marked approximately by the seashore.

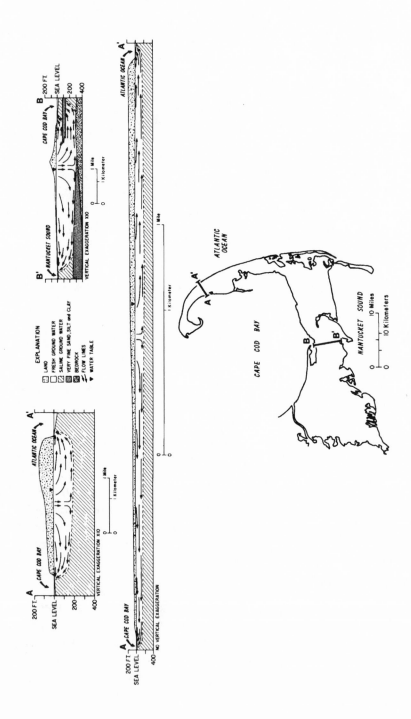

140

In an unconfined aquifer, like sandy outwash deposits, the top of the ground water or water table is indicated by the altitude of the surface of lakes and ponds. For example, in the Mashpee pitted outwash plain, the water level in the ponds and lakes ranges from near sea level along the shore to nearly 70 feet above sea level at the apex of the plain in the town of Sandwich. The depth to the bottom of an aquifer that is entirely within unconsolidated deposits is generally 40 times the altitude of the water table. The bedrock under the Cape and Islands is essentially impermeable and is not considered to be part of the aquifer. In areas where the bedrock is shallower than 40 times the height of the water table, the entire unconsolidated section below the water table will be saturated with freshwater. However, in most places on the Islands and outer Cape Cod, the bedrock surface is well below the aquifer. For example, in the deep borehole on Nantucket, the base of the freshwater was reached at about 520 feet and bedrock was encountered at 1,542 feet.

The hydrologic conditions beneath Nantucket are, however, more complex than freshwater above saltwater. Two zones of freshwater occur within the salt groundwater. One zone occurs

Figure 74. (opposite)   Hydrologic cross sections of Cape Cod. The upper right cross section (B-B') is for the town of Barnstable and depicts the hydrology of inner Cape Cod from Cape Cod Bay to Nantucket Sound. The aquifer is restricted by bedrock and impermeable fine-grained deposits. A perched aquifer occurs above a lens of impermeable sand, silt, and clay beneath the Cape Cod Bay shore. Under this condition, an offshore well could obtain freshwater from below the clayey layer. The upper left cross section (A-A') is for the town of Truro on the lower Cape. Here, the aquifer is unrestricted and is underlain by saline ground water. Vertical exaggeration of the upper cross section is ten. The lower profile shows the Truro cross section (A-A') with no vertical exaggeration. Hydrologic cross sections from the U.S. Geologic Survey Hydrologic Atlas HA-692, compiled by Denis R. LeBlank and others, 1986.

from 730 to 820 feet and a second zone occurs from 900 to 930 feet. The water in these zones is trapped between impermeable clay layers and may be a remnant of freshwater recharge that occurred during glacial stages when the continental shelf was subaerially exposed.

Ground water, like surface water, flows downhill. The flow is nearly horizontal and radial from higher areas to lower areas of the water table. Eventually, the freshwater reaches the seashore, the lowest point on the water table, and is discharged to the ocean. The rate of ground-water flow is largely controlled by the porosity and permeability of the unconsolidated deposits, which in turn, are controlled by the texture of the deposit. In sand and gravel, the rate of flow is on the order of several feet per day; in silt and clay, it may be considerably less than 1 foot per day.

Although the hydrology of the Cape and Islands appears to be simple in a generalized picture, the flow of ground water is not everywhere uniform. In places, the subsurface materials are complex, consisting of strata of differing porosity and permeability. This condition occurs mostly in the moraine deposits and in the glacial-lake deposits, where clay-rich strata are interbedded with layers of sand and gravel. The clay-rich strata may stop or greatly slow the movement of ground water. In places, a local water table may form above a clayey layer of low permeability and porosity. Such local water tables are said to be perched. In these areas, the altitudes of lakes and ponds may not indicate the level of the regional water table. For example, Deep Pond in Falmouth, located in the Buzzards Bay moraine, is more than 50 feet above the actual water table. Perched water tables provide an unreliable source of ground water that may fail during drought or through pumping.

Because the unconsolidated deposits on the Cape and the Islands are very thick, they can store vast amounts of freshwa-

ter. Arguments for the restriction of development and population of the Cape and Islands that are based on the assumption
that the ground-water resource is limited usually fail. A better
argument is one based on the amount of recharge needed to
maintain the aquifer. If the amount of recharge is exceeded by
the amount of water removed from the aquifer, the ground
water is being mined. Under this condition, the water supply
will diminish. The most dramatic result of mining ground
water is a fall in the water table, causing a corresponding fall in
the levels of lakes and ponds and in shallow water wells. A less
obvious result is that the aquifer becomes thinner. For each one
foot drop in the water table, the base of the aquifer rises 40 feet.
This thinning of the aquifer is unnoticeable until wells run dry
or begin to draw saltwater.

A much more serious threat to the ground-water resource is
contamination. Potential sources of ground-water contamination are numerous and include point sources such as dumping
grounds, buried fuel storage tanks, stored road salt, leachate
from sanitary landfills, and municipal waste-water disposal
sites. Dispersed sources of ground-water contamination include highway salt, fertilizers, and domestic septic systems.
Point sources of pollution contaminate a plume of ground
water that may make domestic and municipal water wells unfit
for use when it reaches them. The casual visitor to the Cape and
Islands is unaware that, in places, the ground water is seriously
polluted. The worst pollution occurs in and around the mili-
tary base in western upper Cape Cod. For years, without
thought to the eventual consequences, occupants of the base
disposed of millions of gallons of pollutants by putting them in
the ground or into the base sewage-disposal plant. The base is
near the highest point on the Mashpee pitted plain and above
the highest part of the water table. Pollutants from the base
entered the water table and, as plumes of contaminated ground
water, began their journey to the sea. On their way, the plumes

have contaminated and made useless municipal and private wells and will continue to do so until all of the plumes have completed their passage to the sea. The passage of the plumes may take as long as 100 years. Thus, the people now on Cape Cod and future generations to come will inherit a serious threat to their water supply as a result of the unthinking acts of earlier occupants of the Cape. The people on the Cape and Islands and those to come must learn from these consequences and do all that they can to protect the ground water—the only source of water—or, like the Ancient Mariner, they will have "Water, water everywhere,/ Nor any drop to drink."

The more dispersed sources of contamination cause a widespread deterioration of water quality. Nutrients from fertilized lawns, farm land, and golf courses and phosphates from leaking septic tanks either run off as surface water or percolate downward to the ground water to enrich the ponds and lakes, causing an increase in the amount of aquatic plant life. As these plants die and decay, they may remove enough oxygen from the water so that fish die. If these sources of pollution are reduced or removed, the freshwater quality will slowly improve as the runoff of polluted surface water is reduced and the excess nutrients in the lakes and ponds disperse and as the contaminated ground water moves slowly by and is eventually discharged to the sea. However, the pollutants carried to the sea by the flow of surface water and ground water will affect the marine environment. Excess nutrients in sea water cause the depletion of oxygen, which, if sufficient, will kill off animal life. Toxins introduced to the seawater enter the food chain and make many marine resources unfit for use. Just as it will take a long time for the pollutants to leave our ground water, it will take a long time for the pollutants to reach the sea and be flushed from the bays, estuaries, and marshes.

Water wells close to the sea can become contaminated by saltwater intrusion. If a large amount of water is pumped from the well, the surface of the water table is drawn down, forming a cone of depression. If this cone reaches the base of the aquifer, the water will first become brackish; if pumping is continued, saltwater will enter the well. In 1908, Provincetown located its well field in Truro in order to avoid water with a high iron content caused by organic deposits beneath the Provincetown Spit. However, as the population increased, especially during the summer, more and more water was pumped and the wells were eventually contaminated by saltwater. Reduced pumping brought the water back to potable levels. In 1977 gasoline was discovered in the water, the result of a leaking fuel storage tank. As a result, Provincetown lost 60 percent of its water supply. Attempts to correct the gasoline contamination have included pumping to remove the contaminated water and treatment to remove the gasoline.

The Provincetown experience shows another way to maintain the quality of the Cape and Islands water supplies. Contaminated ground water can be cleaned and made usable as a water supply. Municipalities along many rivers do just that. Each city or town takes water from the river, cleans it up, uses it, and returns the water to the river. New Orleans, near the mouth of the Mississippi, uses river water for its water supply. If the very much used Mississippi River water can be cleaned and used once more, then water from contaminated aquifers can certainly be cleaned too. I hope that the Cape and Islands aquifers will never be used in such a manner, because a water source that needs to be cleaned and decontaminated before it can be used is very expensive. When newly contaminated after use, water should be cleaned again before being returned to groundwater and the ponds, rivers, and saltwater bodies, an equally expensive process.

General information on water resources and specific information on the ground water resources of the Cape and Islands are included in Hydrologic Investigations Atlases published by the USGS (Appendix D).

QUATERNARY HISTORY OF THE CAPE AND ISLANDS

If the geological history represented by the bedrock and coastal plain deposits is considered, the development of the Cape and Islands spans many hundreds of millions of years. However, the geological record contained in these older rocks is incidental to the landscape and will not be dealt with further. Instead, I discuss only the events that shaped the Cape and Islands. Thus, the geologic history here is restricted to the Wisconsinan glacial stage and the Holocene Epoch. During the Wisconsinan glacial stage, the Laurentide ice sheet, the last of the great continental glaciers in North America, advanced to and retreated from the Cape and Islands. During the Holocene Epoch, which may be only an interglacial stage, sea level rose to drown the region. These events in the geologic history are the ones that shaped the Cape and Islands.

As world climate cooled at the end of the Sangamonian interglacial stage, about 75,000 years ago, the Laurentide ice sheet began to form in Canada. During the early Wisconsinan, the ice sheet remained in Canada or made limited advances into the northeastern United States. In middle Wisconsinan time, the climate warmed somewhat, although it was still cooler than today, and the Laurentide ice sheet retreated. During these intervals, the pre-Wisconsinan Cape and Islands, if they existed at all, may have been left high and dry as sea level fell to levels well below present sea level. During late Wisconsinan time, the climate cooled again, and about 25,000 years ago, the Laurentide ice sheet began its advance toward southern New England.

Minimum radiocarbon ages on material found in the glacial drift indicate that the ice reached southern New England, including the Cape and Islands, about 21,000 years ago. When the ice stood at the approximate limit of its advance, it formed the moraines and outwash plains of Martha's Vineyard and Nantucket. South of the ice margin, meltwater streams and rivers flowed across the coastal plain to the seashore, which was close to the edge of the continental shelf. General retreat from the Islands began before 18,000 years ago as the climate began to warm. Glacial lakes formed in front of the ice front and the lakes increased in size as the ice retreated northward. When the ice front reached a position near the eastern side of Buzzards Bay and the southern part of Cape Cod Bay, the retreat stopped and the ice stagnated. Meltwater streams from the Buzzards Bay and Cape Cod Bay lobes built the large outwash plains of western Cape Cod. Shortly thereafter, readvances of the ice front formed the Buzzards Bay and the Sandwich moraines. The Harwich outwash plain was formed when the eastern part of the Cape Cod Bay lobe again stagnated. After the Harwich outwash was deposited, the ice front retreated from upper Cape Cod, and Glacial Lake Cape Cod developed in the southern part of Cape Cod Bay. Other readvances of the two western lobes are represented by a submerged end moraine beneath Billingsgate Shoal and by end moraines north of the Cape Cod Canal. To the east, the South

Channel lobe began to build the series of outwash plains of the lower Cape, most of which were deltas built into Glacial Lake Cape Cod. Further retreat of the Cape Cod Bay and South Channel lobes drained the lake. During deglaciation, the valleys that cut the outwash plains were formed, possibly by spring sapping. The kettles were formed as ice blocks melted. Winds deposited the eolian layer atop the glacial drift. Retreat of the ice into eastern Massachusetts and into the Gulf of Maine ended the glacial development of the Cape and Islands.

Except for the formation of the kettles over slowly melting blocks of buried ice, all of these late Wisconsinan events happened very quickly, perhaps during a thousand years or so. Major glacial features such as the end moraines and outwash plains may have formed in a few hundred years.

The age of the oldest organic deposits on the glacial drift surface indicates that the glacial landscape may have remained essentially unvegetated for several thousand years. Throughout the Wisconsinan, the coastal plain south of the glacial limit had a cold periglacial climate, and vegetation was tundralike, with low bushes, grasses, and stands of arctic trees. Nevertheless, this arduous environment was the place of refuge for plants and animals that would first colonize the land made available by the retreat of the Laurentide ice sheet (Fig. 75). The plants and animals that presently characterize southern New England retreated far to the south during the Wisconsinan stage and returned very slowly following glacial retreat. Many tree species did not arrive in southern New England until the early Holocene and some of the trees that are common in the present forests of southern New England, including the Cape and Islands, arrived during the middle Holocene.

It appears from a recent discovery on the lower Cape that Indians may have arrived in the Cape and Islands region

Figure 75.   Among the plants and animals that occupied the exposed continental shelf south of the Laurentide ice sheet were mastodons. Mastodons and many other animals and plants migrated northward following the retreating Laurentide ice to occupy the Cape and Islands region. This mastodon and calf are part of a life-size diorama at the New York State Museum in Albany. Photograph supplied by Robert H. Fakundiny, New York State Geologist.

between 11,000 and 8,000 years ago. This early occupation is entirely possible, because evidence from elsewhere in New England indicates the presence of Paleoindians by 11,000 years ago. In the fall of 1990, waves from the first northeast storm of the season eroded the cliff face at Coast Guard Beach in Eastham on the lower Cape. The cliff erosion exposed an archaeological site under the coastal dunes. The dunes had protected the site from plowing, which would have mixed up the stratigraphy, and had hidden it from pot hunters, who might have scavenged the site with little regard for its scientific importance. Because the site was threatened by storm waves, which were expected throughout the winter months, an intensive effort was made to excavate and preserve the archaeological material it contained. The material included stone tools and flakes (from the manufacture of stone tools), pottery fragments, and evidence of a hearth and dwelling. Although a preliminary interpretation of the evidence suggests that the site is from 6,000 to 8,000 years old, it may turn out to be considerably older, possibly 11,000 years. Charcoal was obtained, which will provide radiocarbon dates to establish the age of the site. Whenever they arrived, these early inhabitants encountered a much different landscape than we know today because sea level was at least 50 feet below present sea level and the shore would have been at least a mile to the east. Thus, these Indians and later Indians were to witness the last great event in the development of the Cape and Islands: the invasion of the sea.

Submergence of the glacial Cape and Islands resulted in great changes in the landscape. The most dramatic event occurred when the islands became separate entities as the bays and sounds were flooded. The coastline was smoothed when headlands that had been left by the glacier were eroded back under the attack of the waves. Longshore drift formed the bars and spits to close off reentrants in the shoreline. Marine scarps developed along the shore, and sand dunes formed on the barrier spits. The enclosed embayments became the sites for salt marshes and tidal flats. As the land was submerged, the sea redistributed the glacial sand to form the shoals and offshore bars, which would later affect navigation and influence the colonial settlement of southeastern Massachusetts.

As sea level continued to rise, there were frequent rapid changes in the shoreline. Waves generated by major storms altered the shoreline, rapidly eroding back the sea cliffs, and changed the configuration of the barrier spits and islands. Storm waves attacked the barriers, eroding new inlets in some places and closing old inlets in other places. Storm waves washed over the barrier, carrying away sand dunes. Beach and dune sediments were washed into the bays and lagoons, causing them to shoal. Offshore, the waves altered the submerged bars and shoals and in places even built islands where none existed before. Although the rising sea has eroded away large amounts of the Cape and Islands, it has also added new lands.

All of the changes in the landscape are not the result of storms and storm waves. Each day, under normal conditions, longshore drift carries sand down the beach to nourish and lengthen the barrier spits. Tidal currents erode and carry debris into the embayments and build the marsh surface upward and shoreward. Prevailing onshore winds erode and deposit sand to change the shape and position of sand dunes. Together, during storms and fair weather, marine erosion, marine deposition, and the movement of sand by the winds have modified the glacial Cape and Islands and produced the landscape of today.

In addition to shaping the Cape and Islands, the rising sea brought with it an abundant food supply for the human inhabitants. Shell heaps or middens occur in many places, indicating the importance of this food source to the Indians. Finfish and marine mammals were also important to the Indians, as were the water fowl that inhabited the bays, salt marshes, and ponds. Today, the maritime environment provides the temperate maritime climate that makes the weather pleasant for most of the year and continues to shape the land to perpetuate the attractive landscape favored by residents and visitors alike.

EARTHQUAKES

Every few years, and sometimes more frequently, the people on the Cape and Islands feel the ground move beneath them. The ground motion is the result of an earthquake somewhere in northeastern North America. Unlike earthquakes in the western part of the United States, those in the East are not associated with surface faulting, but occur along diffuse zones that may be related to deeply buried fault zones. The western earthquakes are more easily understood, because they occur along the junction of two or more colliding crustal plates. The earthquakes in the eastern part of the United States occur within a crustal block or along its trailing edge and are caused by factors that are little understood. Some of those in the glaciated regions may be related to crustal depression and rise caused by the advance and retreat of the Laurentide ice sheet. Others may be a legacy of the opening of the Atlantic during the Mesozoic Era, acting along very ancient zones of crustal weakness. Eastern earthquakes can be as great as those in the West and are generally felt over a much wider area. In fact, the eastern part of the United States may hold the record for the strongest earthquake. Three quakes in New Madrid, Missouri, in the early 1800's were felt as far east as Boston. Fortunately, earthquakes in the eastern part of the United States occur much less frequently than on the West Coast. Can a very large earthquake hit the Cape and Islands? The answer to that question lies in the earthquake history of the region.

To understand this history, a scale is necessary to measure and compare the relative size of earthquakes. One such scale is called the Modified Mercalli scale (Appendix E). The scale is based on the earthquake damage to manmade structures, damage to the ground surface, and the human reaction to the earthquake. The scale has twelve intensity values. The lowest value indicates an earthquake felt by a few people under favorable circumstances, and the highest value indicates a quake that caused total or nearly total destruction to manmade structures, visible waves in the ground surface, and objects to be thrown into the air. Because the Mercalli scale is based on direct observations during and after the earthquake, it can be applied to establish the intensity of past as well as recent earthquakes. The intensity of the 1989 World Series (Loma Prieta, California) earthquake is placed at 7 to 9 and provides a basis of comparison for large historical earthquakes that have occurred on or near the Cape and Islands.

The recorded history of earthquakes in New England goes as far back as the Pilgrims. In 1638, the Plymouth Colony experienced a very large earthquake, with an estimated intensity of 9. Without doubt, that earthquake affected the Cape and Islands. In 1755, an earthquake with an intensity of 8 occurred near Cape Ann, Massachusetts, and was probably strongly felt on the Cape and Islands. These two large earthquakes occurring at or near the Cape and Islands are sufficient reason to expect that an earthquake with an intensity of 7 to 9 may occur in the future. It has been estimated that a major earthquake will occur somewhere in the eastern United States within the next 20 years.

Damage from a major earthquake on the Cape and Islands would likely be highly variable, and in some places it could be severe. The earthquake shaking could result in the liquefaction of water-saturated sand, silty sand, and clay that would remove the support of buildings. Most wooden homes would likely survive with only minor damage. Masonry buildings, particularly multi-storied buildings that lack steel reinforcement designed to withstand horizontal acceleration, could be severely damaged.

COASTAL STORMS

Each year during late summer and early fall, the people of the Cape and Islands watch with intense interest the development and track of Atlantic tropical storms. When the storm is predicted to pass close to the region and has reached, or is predicted to reach, hurricane force, preparations are made to protect life and property. Although winter northeast storms may be nearly as powerful, the anxiety and storm preparations are usually less intense because at that time the summer visitors have left, the summer homes are buttoned up, and most of the boats are safely stored on land.

Both hurricanes and northeast storms generally approach the Cape and Islands from the south. All storms in the Northern Hemisphere rotate counter-clockwise. The wind speeds on the right side of the storm, as one faces the direction the storm is moving, are increased by the storm's forward motion while the wind speeds of the left side are reduced by this same motion. Because of this, the right side of the storm is called the dangerous semicircle. Winds in the right side of the storm blow from the southeast and southwest and the winds in the left side blow from the northeast and northwest; thus, the speed and direction of storm winds over the Cape and Islands depend on the path of the storm. Storm paths to the west of the region place the Cape and Islands in the dangerous semicircle and produce storm winds from the southeast and southwest; paths to the east produce storm winds with less speed that blow from the northeast and northwest.

For vessels at sea, the most dangerous aspect of storms is the large waves generated by the winds. Ashore, when a storm approaches the coast and moves inland, other aspects become important. Along the coast, extra-high storm tides may flood coastal lowlands. These storm tides are the result of vast amounts of water pushed before the wind that cause a rise in the level of the sea. The rise in the sea surface is also augmented by low barometric pressure that causes an additional rise in water level. If this departure from normal sea level is of short duration and very high, it is called a tidal wave by the local people. It is more appropriately called a storm surge (Fig. 76). When the storm surge hits the coast, it causes a rapid rise in sea level and floods low-lying areas. On top of the storm surge, the high winds build storm waves, which hit the coast and coastal structures with great force. The shape of the shoreline also influences the elevation of the storm surge. Funnel-shaped embayments like Buzzards Bay cause the storm surge to rise in its passage up the bay. This rise causes the greatest flooding in the upper reaches of the funnel-shaped bay. Additionally, storm rains may cause serious stream flooding, and the winds may topple trees and blow down buildings. All of these features are amplified during hurricanes, the most powerful of coastal storms, which thus cause the most damage.

The most famous storm to hit the Cape and Islands was the September 1938 hurricane. A look at its impact will provide an idea of what can be expected when the next major hurricane hits

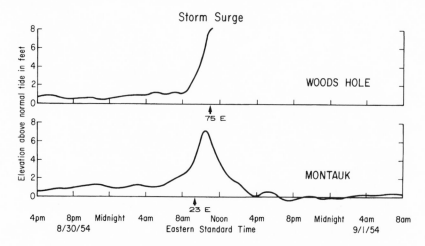

Figure 76. The storm surge recorded by tide gages at Montauk on Long Island and Woods Hole on Cape Cod during Hurricane Carol in 1954. The surge rose about 7 feet in about 3 hours at Woods Hole before the tide gage was carried away. Arrows show the distance in nautical miles east of the storm center. Modified from an illustration by Alfred C. Redfield and Arthur R. Miller of the Woods Hole Oceanographic Institution.

the region. This tropical storm was first noted on September 13th near the Cape Verde Islands, off the coast of Africa. It began to move westward. On the 18th, the storm was a full hurricane and had entered the western Atlantic about 400 miles east northeast of Puerto Rico. By the afternoon of the 20th, the hurricane was east of Jacksonville, Florida, and was becoming trapped between high pressure over the east-central United States and a high over Bermuda. On the morning of the 21st, the day the great storm hit the Cape and Islands, the storm was east of Norfolk, Virginia, and was moving northward at about 70 miles per hour. Shortly thereafter, the winds began increasing off the New England Coast and at 2 p.m. the wind reached hurricane force, more than 74 miles per hour, at Block Island. By 3 p.m., the eye of the storm crossed the south shore of Long Island at about the longitude of Port Jefferson. The track placed the Cape and Islands in the dangerous semicircle of the hurricane. For Rhode Island and southeastern Massachusetts, the hurricane would hit at a time that coincided with the time of high tide.

When the storm arrived at the coast, the forward motion of the 1938 hurricane was 60 to 70 miles per hour. The winds that hit the Cape and Islands ranged from 75 to 90 miles per hour, and short-term peak gusts were close to 100 miles per hour. The great winds toppled many trees and blew away some roofs, but for the most part, direct wind damage was not the greatest threat to the Cape and Islands. The storm surge and the accompanying storm waves caused the most loss of life and property damage, especially at Menemsha, the lowland that separates the town of Gay Head from the rest of Martha's Vineyard, and along the eastern shore of Buzzards Bay.

A book written by Everett S. Allen, entitled *A Wind to Shake the World* (1976, Little, Brown, and Company), describes the effect of the 1938 hurricane on residents of the coast from New Jersey through New England. From Squibnocket Point on Martha's Vineyard, an eyewitness of the storm surge saw a wave with a 25-foot-high crest, with spray and foam 10 feet above that; it was followed by several waves of lesser height. In Chilmark, also on Martha's Vineyard, the storm surge was experienced by some summer residents who lived in a house behind the coastal dunes. At first only sea foam topped the dune. Then small streams of solid water began to run down the landward face of the dune. Two shallow streams of water flowed past the house. As they watched the storm, the water depth began to increase, and they decided to leave. At that time, the water around the house was less than 2 feet deep. After they walked a few tens of feet, a large wave rushed inland and the water was up to their necks. In a half-minute, a second wave brought the water over their heads, and they had to swim for their lives.

In Buzzards Bay, the high tide and storm surge coincided. In the upper reaches of the bay, the storm surge was 13 feet above the predicted level of the high tide. Initially the storm surge, driven by southeast winds, brought high water levels and damaging storm waves to the west side of Buzzards Bay. Shortly thereafter, the hurricane wind shifted to the southwest, and the surge rushed across the bay to hit the eastern shore. The water level rose rapidly until it was about 16 feet above mean low water at West Falmouth and almost 19 feet above mean low water at Bourne. At Wing's Neck, at the west entrance to the Cape Cod Canal, waves atop the storm surge were 8 feet tall and tossed rocks through windows to a height of 29 feet above mean high water. In New Silver Beach, numerous houses were located on a low barrier spit that had been much modified by leveling the dunes and filling the lagoon. The storm surge was nearly 16 feet above mean low water and large waves driven by the southwest wind topped the surge. Most of the houses on the beach were damaged greatly, and quite a few washed away completely (Fig. 77). In a few hours, it was all over and sea level returned to normal. After the storm, the town of Falmouth reported $500,000 worth of property damage, which is the value of a modestly priced waterfront home today.

Can all this happen again? It already has. Hurricane Carol hit the south coast of New England in 1954. The storm was almost a twin of the 1938 hurricane and resulted in great damage along the shore. During both hurricanes, the storm surge and high tide coincided in Buzzards Bay. Because the 1954 storm track was somewhat closer to Buzzards Bay, the storm surge was higher than that in 1938 and reached nearly 16 feet. At New Silver Beach, many of the houses rebuilt after the 1938 hurricane had to be repaired or replaced again. A major hurricane hit the Cape and Islands in 1944. Fortunately, for the residents of the Buzzards Bay shores, the storm surge did not coincide with high tide. Unfortunately, for people living along

Figure 77.   Storm surge and wave damage during the 1938 hurricane along the southeast shore of Buzzards Bay. These are before and after photographs of New Silver Beach in Falmouth. The arrow shows the original position of the only house visible in the lower photograph. The remainder of the houses were washed into the harbor, which can just be seen at the left on the upper photograph. Photographer unknown.

the north shore of Vineyard and Nantucket Sounds, the high tide and storm surge occurred together, and severe property damage was experienced in those places.

During the past few decades, a great many houses have been built in areas historically subject to storm surges during hurricanes and major northeast storms. Many of these houses sit smugly behind sea walls designed to resist normal storm waves and water levels. However, in the past, the sea walls proved useless against the water levels and great waves associated with hurricanes. When the next major hurricane hits the Cape and Islands with a storm surge that coincides with high tide, the cost of the property damage (in 1938 dollars) may be several orders of magnitude greater than that reported by Falmouth for the 1938 hurricane.

Fortunately, we seem to have learned one lesson, and we have the benefits of modern communications so that most people living along the shore watch the progress of hurricanes as they move across the Atlantic and up the coast. They know enough to leave their homes and seek shelter inland when a hurricane warning is issued and remain away until the storm has passed. They can pick up the pieces when they return. In this way, major loss of life is prevented. Once when the author (whose house was at New Silver Beach) was at sea and a hurricane was predicted to hit the Cape and Islands he sent a radio message to his wife. The message said "lock up the house and take the kids and cats inland." My wife asked the man who was relaying the message by telephone if there was anything further. He answered "I presume love." Even though warned, some people may stay at the shore to watch the storm, and lives may be lost. Will the people learn their lesson and decide not to build their homes in places exposed to storm waves and surges. Probably not, but there is some hope as laws are now in place to discourage such practices.

The question as to when the next hurricane will hit the Cape and Islands has now been answered. On August 19, 1991, Hurricane Bob, which had formed a day or so earlier off the southeast coast of the United States, made landfall just to the west of the Cape and Islands. Although it was rated as a category 3 storm, in many places, it was nowhere near as destructive to waterfront property as the 1938 and 1954 storms. Nevertheless, Hurricane Bob caused over a billion dollars in damage. One lesson has been learned. Few people stayed in structures threatened by the storm surge, and the loss of life was very low. One benefit of Hurricane Bob is that there is a new generation that now knows what a hurricane can do. One problem is that those who experienced Bob may be complacent and in the future may ignore warnings and be caught unprepared by a much bigger storm. Some people may believe that since we have had Hurricane Bob the Cape and Islands will be safe for another decade or two. This is untrue, because each year many tropical storms occur that can become hurricanes, and their paths are random. Thus, in any given year, we could have an experience similar to the one that occurred in 1954. In that year, the New England region was hit by Hurricane Carol in August, by Hurricane Edna in September, and by Hurricane Hazel in October!

THE TIDES

The tides provide a rhythm that regulates the lives of many inhabitants of the Cape and Islands. The shellfisherman needs to know the times of low tide whereas the fin fisherman may plan work according to the tidal cycle. Boaters need to know the state of the tide and the strength and direction of tidal currents. Even people who go to the beach may plan their day around the tides, especially in Cape Cod Bay, where the low-tide shoreline may be far seaward of the high-tide shore.

Generally, the tides around the Cape and Islands are semi-diurnal. There are usually two high tides and two low tides during the 24-hour day. The time between high and low tides is about six and a quarter hours. The combination of two high and two low tides requires 24 hours and 50 minutes, and so the tides occur about 50 minutes later each day.

The tides are mostly caused by the gravitational attraction of the moon and to a lesser degree the sun. Roughly twice each month the moon and sun are in line, either on the same side of the Earth (a full moon) or on opposite sides of the Earth (a new moon). At these times, the high tide is higher and the low tide is lower than at other times of the month; the full-moon tide is the highest of all because the sun and moon are on the same side of the Earth. These twice-monthly highest and lowest tides are called spring tides. Twice a month, the moon and sun are at right angles to each other, and the range (the vertical distance between high and low tide) is the smallest. These minimal tides are called neap tides.

Tides around the Cape and Islands are complex; in general, they are the result of the water flooding northward across the continental shelf and ebbing southward. The simplest tides occur in Buzzards Bay and Cape Cod Bay. In Vineyard and Nantucket Sounds, the tides are complex because the tides enter and leave these water bodies by three passages: on the west, via Vineyard Sound; on the east, via the opening between Monomoy Island and Great Point on Nantucket; and on the south, via Muskeget Channel between Martha's Vineyard and

Nantucket. The tides that enter the Gulf of Maine and eventually Cape Cod Bay flood across Georges Bank and through Great South Channel and Northeast Channel.

Tide ranges vary considerably around the Cape and Islands. In Buzzards Bay, the average tide range is about 4 feet. The time of high tide is about 30 minutes later at the upper end of the bay than at the bay mouth. In Vineyard Sound, the tide range averages about 1.5 feet; in Nantucket Sound, the tide range averages about 2.5 feet. Along the eastern shore of Nantucket and Cape Cod, the tide range is roughly 5 feet, and, in Cape Cod Bay, it is closer to 10 feet. In broad, unrestricted parts of the bays and sounds, the currents caused by the flood and ebb of the tides are weak, generally only a fraction of a knot. Where the tidal flow is restricted, for example in Vineyard Sound between the Martha's Vineyard shore and Middleground Shoal, tidal velocities can reach a few knots.

Very strong hydraulic currents occur in a few places. These currents are caused more by a difference in tidal water level between two points than by the ebb and flood of the tide. Such currents occur in restricted passages such as the holes between the Elizabeth Islands, which connect Buzzards Bay with Vineyard Sound, and in the Cape Cod Canal, which connects Cape Cod Bay and Buzzards Bay. Hydraulic currents in Woods Hole reach 7 knots during spring tides. In the Cape Cod Canal, hydraulic currents are generated by a difference in the tidal ranges and a difference in the state of the tide at the two ends. Tides vary by 4 feet in Buzzards Bay and by about 9 feet in Cape Cod Bay, as well as by a 3-hour difference in the state of the tide. Currents in the canal can reach 5 knots. These strong currents were a concern to the early designers of the canal, and some thought was given to placing locks at either end. However, the designers decided that the currents were an advantage because they would keep the canal open throughout the winter.

Strong tidal currents and hydraulic currents are capable of moving great amounts of sand. In some places, such as the holes between the Elizabeth Islands, they scour the bottom and prevent shoaling. On the other hand, they move sand into the Cape Cod Canal and induce shoaling, so that, from time to time, the canal must be dredged to maintain the channel depth. In the new break through the barrier off Chatham, tidal currents are filling the lagoon with sand. In a short time, the lagoon has been almost completely filled just south of the break (Fig. 64, lower). Tidal currents create the sand waves on the Brewster sand flats (Fig. 55), and they move sand on the shoals in Nantucket and Vineyard Sounds. However, on the sand flats and shoals, the net transport in any one direction is small because of the oscillatory nature of the tidal currents.

THE GEOLOGY AND MAN

The glacial drift supplied the Indians with quartz and volcanic rocks for tools. It also supplied abundant freshwater, forests for fuel and building materials, gentle topography and waterways for easy travel, an easily tillable soil, and clay for pottery. Each spring, streams flowing from freshwater ponds and lakes to the sea were full of herring, a source of food and fertilizer. The barriers and baymouth bars provided the Indians with sheltered embayments in which to fish and hunt fowl. Tidal flats supplied abundant shellfish and, at times, stranded whales, dolphins, and turtles.

For the colonials and later inhabitants, the drift and marine deposits provided all these things and more. There were boulders for stone walls and house foundations and clays from glacial-lake deposits for bricks. Clay in the Gay Head Cliff was considered to be of a high quality and was mined and shipped to Boston for the making of pottery. The abundance of sand on Cape Cod attracted the glass industry, but the sand proved to be unsuitable because of the iron it contained (the sand used to make the famous Sandwich Glass was imported to the Cape). Freshwater marshes and swamps furnished the early inhabitants with peat and bog iron ore.

The freshwater marshes and swamps also became the sites for growing cranberries, which today is the most important agricultural cash crop on the Cape and Islands. The well-drained flatlands of the outwash plains encouraged agriculture during the nineteenth century and early twentieth century, when asparagus and strawberries were among the important crops. Even before that, the Cape and Islands produced grain for New England.

Well-sheltered harbors encouraged the fishing and the whaling industry. The streams and tides supplied waterpower for mills. In the sun, seawater was evaporated to make salt to preserve fish for market and for export. Even the offshore bars contributed to the Cape and Islands economy, by trapping ships for the wreckers. Today, the income of many residents and seasonal workers, as well as the overall economy of the region, is heavily dependent on tourism and real estate. In turn, these two factors are dependent on the preservation of the unique maritime character of the Cape and Islands.

Other features of the geology have influenced man in different ways. The shoals south of Nantucket (Fig. 54) turned the Pilgrims away from their intended destination and caused them to settle in southeastern Massachusetts. The hazard to navigation presented by the shoals and offshore bars along the Atlantic side of the Cape and Islands was one of the primary reason for the building of the Cape Cod Canal. Another reason was to avoid naval blockades and enemy submarines. German submarines operated close to the Cape and Islands during World Wars I and II. Good harbors determined where the villages would be on the Cape and Islands. The outwash plains were the best sites for farms while the moraines were left as woodland and had little value as real estate. In 1888, Nathaniel Shaler, a geologist, noted that land in the Martha's Vineyard moraine was worth about $2 an acre and could probably be bought for less, a far cry from the present value of such land!

If man has been influenced by the geology of the Cape and Islands, he has also been an agent in the geology. Through some of his activities, natural processes are influenced and the landscape is modified. Sand and gravel pits may be an eyesore to many, but they can be a delight to geologists because they provide exposures that help him to understand the nature and origin of the glacial deposits. Power lines, gas pipelines, and public works, such as erosion control, harbor maintenance, schools and hospitals, roads, water and sewer facilities, are all necessary, but they may cause important changes to the environment and landscape. Therefore, it is equally necessary to understand the environmental impact when such structures are planned and built.

Efforts to control the erosion of the sea cliffs and changes in the barrier islands through the construction of sea walls and jetties and hardening or filling breaches in the barrier beach system may do more harm than good (Fig. 78). Houses or condominiums built on coastal land come with risks that must be recognized and accepted by the property owner. Arguments for efforts to protect structures on land subject to flooding and

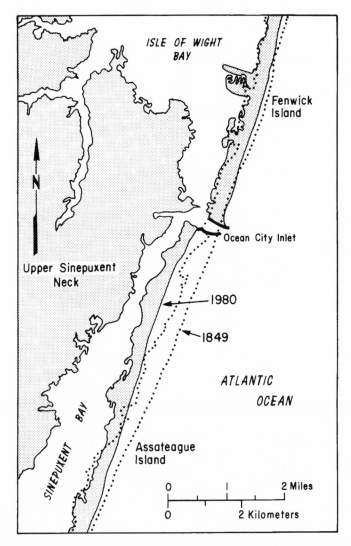

Figure 78.   One result of man trying to stabilize a natural break through a barrier island. This example comes from Ocean City, Maryland. Assateague Island has migrated landward relative to Fenwick Island as a result of the stone jetties placed to control the migration of Ocean City inlet. A similar condition might result if the Chatham break is stabilized by a hard structure, and if sand is no longer transported along the shore to nourish the barrier to the south.

wave attack during northeast storms or hurricanes should be carefully weighed. In the long run, these structures are certain to suffer damage or be lost completely; in the meantime, efforts to protect them may cause problems elsewhere along the shore. For example, a sea wall built to stop shore erosion may remove the source of sand that naturally would go to nourish the beach further along the coast. Under these conditions, the beach may narrow or erode away completely, so that the protection it offered the upland is lost.

The quality of the water supply can easily and very quickly be degraded. Leaking fuel tanks, inadequate handling of septic waste and hazardous materials by private and public facilities, uncontrolled addition of nutrients, including fertilizers, and so-called sanitary landfills, which contain unintentional or intentional concentrations of hazardous wastes, are all serious threats to the water supply and to the water quality of our lakes, ponds, and bays.

For the Cape and Islands, the optimum time to recognize man's adverse effect on the natural landscape and water supply may already have passed. However, identification of the sources of environmental harm and efforts to correct them may do much to restore the environment. In addition, much can be done now and in the future to preserve as much of the natural landscape as possible, given that the Cape and Islands will continue to be developed and that the need for changes and additions to the infrastructure will have to be addressed.

Arthur Strahler published *A Geologist's View of Cape Cod* in 1966 (1988, Parnassus Imprints), an informative and easy-to-understand book on the geology. Professor Strahler came to Cape Cod after he retired from the faculty of Columbia University. After years on the Cape, he moved to Santa Barbara, California, and wrote a short essay entitled *Cape Cod Viewed*

From Santa Barbara (1973, Association for the Preservation of Cape Cod). His observations in this essay are still important to the future of the Cape and Islands. I here excerpt a number of passages from *Cape Cod Viewed From Santa Barbara* in the hope that they will receive attention once again. Here is some of what Arthur Strahler wrote:

> "If it be true that the best way for an American to gain insight into his United States is to travel abroad, perhaps one can gain insight into the workings of Cape Cod by leaving it to live elsewhere. The learning process is particularly intense when one's new home is another coastal locality having many environmental problems in common with the Cape . . .
>
> While the physical setting of the Santa Barbara (County) area may seem strangely different from that of the Cape, there are some basic structural likenesses. Both regions are long, narrow, strips of land with severe limitations on space for development . . .
>
> Perhaps the most striking difference between the two lies in the more advanced state of Santa Barbara's awareness that it has serious problems and that it must solve them or cease to be a desirable place to live or visit. In contrast to the general apathy of Cape Codders to environmental changes that will affect them adversely, citizens of the Santa Barbara area are a large and vocal segment of the population. They are attempting, and in some cases succeeding, to secure enactment of legislation to protect their future. They are putting up for political office candidates who will speak for their principles. They seek substantial revisions of systems of plans and regulations affecting growth of the area. In contrast, Barnstable County, appears as a near non-entity in the exercise of government functions. All of the land of the Cape is under the immediate and direct authority of the towns, each of which believes it can manage its own affairs and direct its own future without any help from its neighbor towns, the county, the state, and even the federal government . . .

A radical alternative to the growth of the Cape along its present lines is offered by conservationists: "Zero growth" or at least a leveling off of growth to reach a manageable level of population, stabilized within tolerable limits of environmental degradation. To preserve what is left of Cape Cod is a lofty ideal, but is it realistic? Preservation faces serious trouble in terms of a deep-seated force—the New England ethic of personal freedom to do whatever one pleases with his own land. Long sanctioned by law and tradition this is the cornerstone of our system of free enterprise. What will the preservationist offer as a substitute to the unbridled exercise of personal freedom? Will he suggest that Cape Codders surrender some measure of it to enhance the public good? . . .

It will be up to the Cape as a single geographical unit to take the initiative to police its own environment and guide its own development. In my opinion, nothing short of a regional authority set up by the voters can save the Cape from transformation into a community unrecognizable as the Cape of today. Whether such regional authority will ever come into existence on Cape Cod is doubtful at best. I suspect many Cape Codders would prefer to go to perdition in their own way, and who am I to tell them otherwise?"

Strahler wrote these words in 1973 and it wasn't until 1990 that a countywide governmental body was formed at the request of the citizens of Cape Cod. The Cape Cod Commission was formed to address the environmental and developmental problems of the Cape. The time lost between Strahlers' warning and the formation of the Cape Cod Commission allowed almost unbridled development with its corresponding degradation of the Cape from what many would wish it to be. On Martha's Vineyard, a commission was created in 1974, but Nantucket, a county in its own right, has no such organization. The impact of man on the Cape and Islands will remain unpredictable, but the future of the region in terms of natural forces may be more easily foreseen.

THE FUTURE CAPE
AND ISLANDS

The future of the Cape and Islands is good news and bad news. The bad news is that the Cape and Islands are slowly being eroded away by the waves. Their ultimate future can likely be seen by observing Georges Bank or Stellwagen Bank, which is located north of the lower Cape. The good news is that sooner or later every piece of property will be water front. If the sea-level rise continues at about its present rate, the Cape and Islands will probably battle the sea for another 5,000 years before they disappear. If the rate of sea-level rise increases, the battle will be shorter, or if it slows the battle will last longer. The first line of defense for the Cape and Islands is the shore—the beaches, barrier islands or spits, and the offshore bars and shoals that absorb the brunt of the wave attack. In places where there is no defense or when the defense is overwhelmed, the upland coast is attacked directly by the waves and retreats rapidly. In other places, the defense is weak or absent, and there the shoreline retreats rapidly each year. Storm overwash, breaches, and flood-tidal deltas are ways that barriers survive. Although breaches and washovers are troubling to many, they are essential to the survival of the barrier as sea level rises. In this way, the barrier rolls over itself, tank-tread fashion, to move landward into shallower water.

Most of the shoreline of the Cape and Islands is undergoing erosion and the shore is retreating (Fig. 79). Even in some

Figure 79.   Erosion of the cliff on the ocean side of Wellfleet during the February 1978 northeast storm. The house just escaped falling down the cliff face, its eventual fate unless it is moved. To the left of the house is part of a crane boom to be used to lift the house to a safer position. Moving the house back from the cliff will increase its life span, but as the erosion continues, at an average rate of about 3 feet per year, the property owner will eventually have no land left.

places where the shoreline is moving seaward, the long-term result is still shoreline retreat because the cliffed and eroding headlands determine the position of the coastline. The rate of retreat varies greatly from place to place. The shoreline exposed to the open ocean retreats fastest. Maximum average retreat ranges from about 7 to 11 feet per year.

Average rates of erosion quoted for the Cape and Islands can be somewhat misleading. During about 30 years at Wasque Point at the southeast corner of Martha's Vineyard, the average erosion rate was about 33 feet a year. During most of that period, the point was protected by a wide foreshore, but one year when the foreshore was absent and the point was exposed to direct wave attack, it retreated 350 feet. Shorelines that face large embayments such as Nantucket Sound and Cape Cod Bay are retreating, but at a much slower rate. The loss of land to the sea by erosion is in no way nearly balanced by the few places where the land surface is growing by accretion. One estimate, for the island of Nantucket, indicates that only about 10 percent of the material eroded from the sea cliffs is carried above sea level in the form of accreting beaches. A somewhat more optimistic estimate, based on the erosion of the great sea cliff of the lower Cape and the growth of the spits to the north and south, indicates that for each acre lost to the sea less than one half an acre is gained by accretion. Nevertheless, eventually the Cape and Islands will lose their battle with the sea, but the sea will not win easily.

The susceptibility of the sandy Cape and Islands to the forces of the sea may be a strength, not a weakness. It could ensure that they will exist in one form or another for a long time. The Cape and Islands may sometime be nothing more than islands of beach sand, capped by dunes and surrounded by sandy shoals. This, however, won't happen for a long time. Maybe we can take comfort and pride that this fragile but adaptable landscape will survive in one form or another long after the rockbound coast of New England and its coastal cities may be deeply submerged by the sea. However, the sand supply is limited, and, if sea level continues to rise, a possible result of the "greenhouse effect" proposed by many scientists, the Cape and Islands will suffer the same fate as Georges Bank and Stellwagen Bank. The fate of Billingsgate Island in Cape Cod Bay off Wellfleet is a likely precursor for all the Cape and Islands as the sea continues to erode the fragile land. In the middle 1800's, Billingsgate Island was about a mile long and about half a mile wide. It was the site for about 30 homes and a schoolhouse, as well as the lighthouse, and supported farming and an important fishing industry. Today it is a shoal exposed only during the very lowest tides (Fig. 80).

Figure 80. The left photograph shows the lighthouse on Billingsgate Island before the entire island was eroded away. It is clear that shore erosion threatens the lighthouse and to the left can be seen a riprap sea wall built in a futile effort to protect the lighthouse from wave attack. Unfortunately, the sea wall was poorly placed and actually increased the rate of erosion. Photograph provided by the Cape Cod National Seashore.

The photograph above shows what remains today of Billingsgate Island off Wellfleet. Photographed at about low tide in 1991. Today all that remains of the island is a shoal exposed during very low tides. The blocks of stone behind the riprap sea wall are all that remain of the lighthouse. The fate of the Billingsgate Island settlement and lighthouse is a precursor for the Cape and Islands as the sea continues to erode the fragile land. Photograph by Dann S. Blackwood, U.S. Geological Survey.

GLOSSARY

ALLUVIAL   Said of a surface graded by streams.

AQUIFER   Permeable body of rock or unconsolidated sediment that yields significant quantities of water for wells and springs.

ARKOSE   A type of sandstone or conglomerate composed of angular fragments of quartz and feldspar. Commonly reddish.

BAR   A ridge parallel to the shore that is emerged at low tide and submerged at least at high tide. Mostly composed of sand.

BARRIER BEACH   Narrow, elongate sand ridge slightly above high water and bordered on the landward side by a lagoon.

BARRIER ISLAND   Long, narrow coastal island with dunes, vegetation, and marshes. Bordered on the landward side by a lagoon.

BARRIER SPIT   Barrier beach or barrier island attached to the mainland at one end.

BASALT   Dark-colored, fine-grained, igneous rock crystallized from lava flows.

BASEMENT   A catchall term for the solid rock that underlies the loose sediments beneath the Cape and Islands.

BAYMOUTH BAR   A beach between headlands that completely or nearly completely closes off an embayment from the open sea.

BEDROCK   Basement.

BUZZARDS BAY LOBE   Lobe of glacial ice that occupied Buzzards Bay and Vineyard Sound during the Laurentide glaciation.

CAPE COD BAY LOBE   Lobe of glacial ice that occupied Cape Cod Bay and Nantucket Sound during the Laurentide glaciation.

CHERT   Rock similar to flint; a noncrystallized variety of quartz, composed of silica.

CLAST   An individual grain or rock fragment.

COASTAL PLAIN   Low, gently sloping, broad alluvial plain between the sea and the uplands.

CONGLOMERATE   Consolidated sedimentary rock composed of rounded to subangular pebbles to boulders.

CONTINENTAL GLACIER   Glacier of considerable thickness that completely covers a major part of a continent.

CONTINENTAL MARGIN   The ocean floor between the shoreline and the deep ocean. Includes the continental shelf, continental slope, and continental rise.

CONTINENTAL SHELF   The sea floor between the shore and the continental slope; the submerged part of the coastal plain.

CONTOUR INTERVAL   Vertical distance between contour lines.

CONTOUR LINE   A line on a map that connects points of equal value. On a topographic map or bathymetric map, a line that connects points of equal altitude above and below sea level.

CROSS STRATIFICATION   Inclined beds within a larger sedimentary layer or lens.

CRUST   The rocky outer layer of the Earth.

CRYSTALLINE   A rock made up of crystals or fragments of crystals.

CUESTA   Ridge with a steep slope on one side and a gentle slope on the other.

DANGEROUS SEMICIRCLE   In the northern hemisphere, the right-hand side of a hurricane as one faces the direction the storm is moving.

DELMARVA   The peninsula between Chesapeake Bay, Delaware Bay, and the Atlantic Ocean made up of Delaware and parts of Maryland and Virginia.

DELTA   Low-lying, flat surface at the mouth of a river, commonly triangular. Underlain by sediments carried to the sea or a lake by the river.

DIORITE Granitelike, coarse-grained crystalline igneous rock containing more feldspar than quartz.

DRIFT General term for all glacial deposits, including those laid down directly by ice and those transported by meltwater.

DROP STONE A stone in lake or marine deposits carried into a lake or the sea by an iceberg.

DRUMLIN Streamlined, elongate hill shaped by advancing glacial ice. Usually composed of till.

END MORAINE A ridge formed at the front of a glacier and underlain by drift.

EOLIAN DEPOSIT Deposit of wind-transported sediment.

EPOCH A block of geologic time within a geologic period, for example, the Pleistocene Epoch.

ERA The primary block of geologic time that includes periods and epochs.

ERRATIC A stone in glacial drift that is different in composition than the underlying bedrock.

EXTRUSIVE ROCKS Igneous rocks formed at or near the Earth's surface.

FELSITE Light-colored, fine-grained, quartz-rich volcanic rock.

FLOWTILL A superglacial till that has flowed onto a stratified drift surface.

FOREDUNE The shore-parallel dune closest to the beach.

FORESET BED Steeply dipping bed deposited at the front of a delta.

FROST WEDGE A vertical wedge-shaped structure formed in permanently frozen ground.

GABBRO A dark-colored, coarse-grained, crystalline igneous rock with little or no quartz.

GASTROLITH Rounded and polished stone from the stomach of an animal.

GEOMORPHOLOGY Study of the shape and origin of landforms.

GEORGES BANK Broad, shallow, elevated sea floor east of Nantucket.

GLACIAL LAKE   A lake impounded in part by glacial ice.

GLACIAL LOBE   Tonguelike projection of an ice sheet.

GLACIAL STAGE   Major subdivision of the Pleistocene when continental glaciers developed.

GLACIOMARINE   Formed in the sea at the front or beyond a glacier.

GLACIOTECTONISM   Deformation of strata by moving glacial ice.

GLAUCONITE   A dull-green iron silicate found in marine sediments.

GLOBAL TECTONISM   Large-scale movements of the Earth, including sea floor spreading and continental drift.

GNEISS   Light-colored, foliated metamorphic rock formed deep in the Earth's crust.

GRANITE   Quartz-rich, coarse-grained, crystalline igneous rock containing abundant feldspar.

GRANITE GNEISS   A gneiss with the composition of a granite.

GRANODIORITE   Coarse-grained, crystalline igneous rock, similar to granite but containing less quartz. Intermediate between granite and diorite.

GREAT SOUTH CHANNEL   Passage into the Gulf of Maine between the Cape and Islands and Georges Bank.

GREEN SAND   Sand containing abundant glauconite.

GROIN   A long, narrow jetty used to trap sand along the beach.

GROUND WATER   Water in the subsurface saturated zone.

GULF OF MAINE   The large coastal embayment between the Cape and Islands and Nova Scotia.

HOLOCENE   The last 10,000 years of geologic time.

IAPETUS   Sea that existed in the general position of the present Atlantic Ocean during the Paleozoic Era.

ICE-CONTACT DRIFT   Drift deposited over and against glacial ice.

ICE-CONTACT HEAD   The collapsed upstream end of an outwash plain.

IGNEOUS ROCKS   Rocks crystallized from molten magma or lava.

ILLINOIAN STAGE   Next to last glacial stage. From about 350,000 to 125,000 years before present.

INTENSITY   Unit on a scale of the relative size of earthquakes.

INTERGLACIAL STAGE   Major subdivision of the Pleistocene when continental glaciers were melted away completely or greatly reduced in size.

INTERLOBATE ANGLE   The angle between two glacial lobes.

INTERLOBATE DEPOSITS   Drift deposited in the interlobate angle between two lobes.

INTERLOBATE REGION   The region between major glacial lobes.

INTERTIDAL   Occurring between high- and low-tide levels.

JETTY   Large structure generally built to protect harbor entrances.

JURASSIC PERIOD   Middle geologic period of the Mesozoic Era, from about 208 to 144 million years before present.

KAME   A knoll or hill underlain by stratified glacial drift that was deposited in a hole in the ice. Larger kames may have a flat, stream-graded surface.

KAME AND KETTLE TERRAIN   A region of hummocky topography made up of kames and kettles. Underlain by collapsed stratified glacial drift.

KETTLE   A topographic depression marking the site of former buried ice block. Usually surrounded by glacial-stream deposits.

KETTLE POND   A pond within a kettle; also a kettle lake.

KNOT   One nautical mile per hour, or 1.15 statute miles per hour.

LAG DEPOSIT   The heavier or larger material that is left behind after finer material has been winnowed away by currents of air or water.

LAGOON   Shallow body of seawater landward of a barrier island, barrier beach, or barrier spit.

LAURENTIDE ICE SHEET   The continental glacier that developed in North America during the last glacial stage.

LAVA   General term for molten rock at the Earth's surface or the solidified rock formed from the molten rock.

LIGNITE   Brownish-black coal intermediate between peat and soft coal.

LIMESTONE   A sedimentary rock that is mostly composed of calcium carbonate.

LONGSHORE CURRENT   A shore-parallel current that results from the oblique approach of waves to the shore.

LONGSHORE DRIFT   The movement of material along the shore, propelled by longshore currents.

LOWER CAPE   Local term for the eastern part of Cape Cod, downwind to the prevailing breezes.

MAGMA   Molten igneous rock beneath the Earth's surface.

MARINE SCARP   A sea cliff.

MARSH   Wet, poorly drained or flooded, generally treeless area with mostly water grasses or grasslike vegetation.

MELTWATER   Water derived from melting glacial ice.

MESOZOIC ERA   Geologic time between about 245 to 66 million years before present; the age of dinosaurs.

METAMORPHIC ROCK   Rock formed mostly deep in the Earth's crust from existing rocks by heat and pressure.

MICROFOSSIL   A microscopic fossil.

MODIFIED MERCALLI SCALE   A scale for determining the intensity of an earthquake. See Appendix E.

MORAINE   A ridge formed of drift deposited along the front or at the sides of a glacial lobe.

MUD   Sediment composed mostly of silt and clay.

OUTWASH   Sedimentary deposit, mostly sand and gravel, deposited beyond a glacial ice front by meltwater streams.

OUTWASH PLAIN   Broad, gently sloping, alluvial surface underlain by outwash sand and gravel that was deposited by meltwater streams.

OXYGEN-ISOTOPE SCALE   Graph showing the relative amounts of two oxygen isotopes (16_0 and 14_0) over time.

OXYGEN-ISOTOPE STAGE   A division of the marine oxygen-isotope scale that generally indicates the duration of glacial and interglacial stages.

PALEOGEOGRAPHIC MAP   A map that depicts the physical geography of a region in the past.

PALEOMAGNETIC   Dealing with the past position of the Earth's magnetic poles and pole reversals.

PALEOSOIL   An old soil that has been buried and that is usually capped by a modern soil.

PALEOZOIC ERA   Geologic time between about 570 and 245 million years before present. The time when life became abundant on Earth.

PAMET   A local name for the outwash plain valleys. Named after the Pamet River valley in Truro on Cape Cod.

PARABOLIC DUNE   A U-shaped dune that opens toward the prevailing wind. Also called a blowout dune.

PENNSYLVANIAN PERIOD   Geologic period of the Paleozoic Era for the time period from 320 to 286 million years ago.

PERCHED WATER TABLE   The top of a saturated layer located above an impermeable layer and above the regional water table.

PERIOD   Primary time division of a geologic era.

PERMAFROST   Permanently frozen ground.

PHYSIOGRAPHIC PROVINCE   A region with similar geology, for example, the Atlantic Coastal Plain or the Appalachian Mountains.

PITTED OUTWASH PLAIN   An outwash plain with abundant kettles.

PLEISTOCENE   An epoch of the Quaternary Period, from about 1.6 million years ago to 10,000 years ago. The Ice Age.

PLIOCENE   The epoch of the Tertiary Period just before the Pleistocene, from about 5.3 to 1.6 million years ago.

PLUTONIC ROCKS Igneous rocks formed deep in the Earth's crust.

PODZOL The major soil group formed beneath a forest in a cool to temperate, moist climate. The major soil of the northeastern United States.

PRECAMBRIAN Geologic time before the Paleozoic Era. The time when life was absent or just beginning on Earth.

QUARTZITE Metamorphic rock formed from fused quartz sandstone, or a sedimentary rock with the quartz grains cemented together by silica.

QUATERNARY The present geologic period, includes the Pleistocene and Holocene Epochs. Usually considered to have begun about 1.6 million years ago.

RADIOMETRIC DATING A method of measuring geologic time by using the rate at which radioactive isotopes decay.

RILL A very small channel. Often used for the very small channels in a beach that carry discharging ground water to the ocean.

SALTATION Transportation by water flow or wind in which the particles move across the streambed or ground surface by a series of jumps or bounces.

SALT MARSH A marsh that is inundated by the high tides, either daily or during the highest tides.

SANGAMONIAN The interglacial stage between the Illinoian and Wisconsinan glacial stages. From about 125,000 to about 75,000 years ago.

SEA CLIFF A scarp cut into upland along the shore by waves.

SEDIMENTARY ROCK A rock composed of grains deposited by water or air.

SEISMIC STUDIES Study of both the deep and shallow structure of the Earth, using earthquake-generated or artificially generated vibrations.

SILICIFIED WOOD Fossil wood in which the wood cells have been replaced by silica.

SOIL HORIZON Layers within a soil.

SOUTH CHANNEL LOBE   A lobe of the Laurentide ice sheet that lay to the east of the Cape and Islands.

SPIT   A long, narrow, sandy point of land constructed by longshore drift.

STELLWAGEN BANK   A submerged bank north of Provincetown and east of Boston.

STORM SURGE   A sudden rise in sea level caused by storm winds and a fall in atmospheric pressure.

STRATA   Layers or beds of sediments.

STRATIFIED GLACIAL DRIFT   Glacial sediments deposited in meltwater streams, glacial lakes, or the sea. Characterized by layering.

TERMINAL MORAINE   An end moraine that marks the maximum advance of a glacier.

TERTIARY PERIOD   The first period in the Cenozoic Era, from about 66 million years before present to about 1.6 million years before present. Beginning of the age of mammals.

TIDAL DELTA   A delta formed by flood- or ebb-tidal currents that transported and deposited sand in breaches through barrier islands or barrier spits.

TILL   Unsorted and mostly unstratified drift laid down directly by a glacier.

TOPSET BEDS   Nearly horizontal stream deposits above the steeply sloping foreset beds in a delta.

TRIASSIC PERIOD   First geologic period of the Mesozoic Era. From about 245 million years before present to 208 million years before present. The time of the first dinosaurs. Beginning of the age of reptiles.

UNCONFORMITY   An erosion surface that represents a time gap during the deposition of geologic strata.

UNCONSOLIDATED ROCK   A layer of sediment composed of loose, uncemented grains.

UNSTRATIFIED DRIFT   Glacial drift lacking layers; a till.

UPPER CAPE   Local name for the upper arm of Cape Cod. The upper Cape is upwind of the lower Cape during the prevailing winds.

VARVES   Annual layers of sediment, common in glacial lake deposits.

VENTIFACT   Stone eroded and polished by wind-transported sand and silt.

VOLCANIC ROCK   Igneous rock formed at or near the Earth's surface during volcanic eruptions.

WATER GAP   A pass cut by a river through a mountain or cuesta.

WATER TABLE   The upper surface of the fresh ground water in the saturated zone or aquifer.

WEATHERING   The mechanical and chemical alteration of rock materials by exposure to the atmosphere.

WISCONSINAN   The most recent glacial stage. From about 75,000 years before present to 10,000 years before present.

APPENDICES

APPENDIX A

Topographic Quadrangle Maps of Cape Cod and the Islands

Below are the 26 topographic maps for Cape Cod and the Islands that are available from the U.S. Geological Survey. All of the topographic maps are 7½-minute quadrangles and have a scale of 1:25,000 (about 2.5 inches to the mile) and a contour interval of 10 feet.

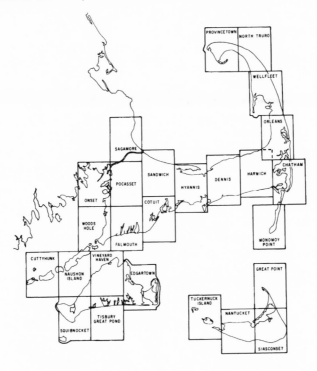

Cape Cod and the Elizabeth Islands

Onset	Pocasset	Dennis
Woods Hole	Falmouth	Harwich
Naushon Island	Sandwich	Orleans
Cuttyhunk	Cotuit	Chatham
Sagamore	Hyannis	Monomoy Point
Wellfleet	North Truro	Provincetown

Martha's Vineyard

Naushon Island	Tisbury Great Pond
Squibnocket	Edgartown
Vineyard Haven	

Nantucket

Tuckernuck Island	Great Point
Nantucket	Siasconset

These topographic maps are available from the U.S. Geological Survey and also may be purchased locally in sporting goods or book stores. Obtaining topographic maps from the U.S. Geological Survey requires two steps. First order an INDEX TO TOPOGRAPHIC MAPS for Massachusetts. This is free. The index will tell you the name and price of the topographic map. (In 1991, each map cost $2.50); this price is subject to change. You can then order it. The map index and maps can be ordered from:

U.S. Geological Survey Map Sales
Federal Center, Box 25286
Denver, CO 80225
(303) 236-7477

APPENDIX B

Map Scales

The following map scales are those of published maps referred to in the book.

Scale	Inches to the mile	1 inch represents
1:100,000	0.63	about 1.6 miles
1:62,500	1.01	about 1 mile
1:48,000	1.32	about .75 mile
1:31,680	2.00	.5 mile
1:25,000	2.53	about .39 mile
1:24,000	2.64	about .38 mile
1:20,000	3.17	about .32 mile

APPENDIX C

Geology and Soils Maps of Cape Cod and the Islands

Geologic Quadrangle (GQ) Maps (scale 1:24,000) Published by the U.S. Geological Survey*

Quadrangle	Map number	Publication date
Sandwich	GQ-1222	1975
Cotuit	GQ-1213	1975
Hyannis	GQ-1158	1974
Dennis	GQ-1114	1974
Harwich	GQ-786	1969
Monomoy Point	GQ-787	1968
Chatham	GQ-911	1970
Orleans	GQ-931	1971
Wellfleet	GQ-750	1968
North Truro	GQ-599	1967

1991 price $3.60; price subject to change.

Other Geologic Maps

Geologic map of Nantucket and nearby Islands, Massachusetts: U.S. Geological Survey Miscellaneous Investigations Series Map I-1580, scale 1:48,000, $3.10.

Geologic map of Cape Cod and the Islands, Massachusetts: U.S. Geological Survey Miscellaneous Investigations Series Map I-1763, scale 1:100,000, $3.

Ordering Information

A 98-page. catalog that includes geologic reports for Massachusetts is sold by the USGS for $1 (1990 price) as "List of U.S. Geological Survey Geologic and Water-Supply Reports and Maps for Massachusetts, Rhode Island, and Connecticut." To order the catalogue or the geologic maps listed above, write check to Department of the Interior-USGS. Mail the order to:

> U.S. Geological Survey Map Sales
> Federal Center, Box 25286
> Denver, CO 80225

Soils Maps

Fletcher, P. C., and Roffinoli, R. J., 1986, Soil survey of Dukes County, Massachusetts: U.S. Soil Conservation Service, scale 1:20,000.

Langlois, K. H., 1979, Soil survey of Nantucket County, Massachusetts: U.S. Soil Conservation Service, scale 1:20,000.

Interim soil survey report, Barnstable County: U.S. Soil Conservation Service, scale 1:25,000. This report is published by the Cape Cod Conservation District. Price $25.

Soils maps published by the U.S. Department of Agriculture are free and may be obtained from:

> U.S. Department of Agriculture
> Soil Conservation Service
> PO Box 709
> Barnstable, MA 02630
> (508) 362-9332

The interim soil survey report for Barnstable County can be purchased from the Cape Cod Conservation District, Flint Rock Road, Barnstable, Massachusetts (price $25) or by mail from:

> Cape Cod Conservation District
> PO Box 296
> West Barnstable, MA 02668

APPENDIX D

Water Resources of Cape Cod
and the Islands

The following reports by the U.S. Geological Survey provide information on water resources.

Delaney, D. F., 1980, Ground-water hydrology of Martha's Vineyard, Massachusetts: U.S. Geological Survey Hydrologic Investigations Atlas HA-618, two sheets, scale 1:48,000, $6.10.

Walker, E. H., 1980, Water Resources of Nantucket Island, Massachusetts: U.S. Geological Survey Hydrologic Investigations Atlas HA-615, 2 sheets, scale 1:48,000, $4.80.

LeBlanc, D. R., Guswa, J. H., Frimpter, M. H., and Londquist, C. J., 1986, Ground-water Resources of Cape Cod, Massachusetts: U.S. Geological Survey Hydrologic Investigations Atlas HA-692, 4 sheets, scale 1:48,000, $14.40.

The 1990 prices above are subject to change. The atlases may be ordered from the following address:

  U. S. Geological Survey Books and Reports Sales
  Federal Center, Box 25425
  Denver, CO 80225
  (303) 236-7476

Other free U.S. Geological Survey publications on water resources:

  A Primer on Water

  A Primer on Ground Water

  Ground Water and the Rural Homeowner

May be obtained from the following address:

  U.S. Geological Survey Books and Reports Sales
  Federal Center, Box 25425
  Denver, CO 80225
  (303) 236-7476

APPENDIX E

Modified Mercalli Earthquake Scale

I   Not felt except by a very few under especially favorable circumstances.

II   Felt only by a few persons at rest, especially on upper floors of buildings. Delicately suspended objects may swing.

III   Felt quite noticeably indoors, especially on upper floors of buildings, but many people do not recognize it as an earthquake. Standing motor cars may rock slightly. Vibration like passing of truck. Duration estimated.

IV   During the day, felt indoors by many, outdoors by few. At night, some awakened. Dishes, windows, doors disturbed; walls make cracking sound. Sensation like heavy truck striking building. Standing motor cars rocked noticeably.

V   Felt by nearly everyone, many awakened. Some dishes, windows, etc., broken; a few instances of cracked plaster; unstable objects overturned. Disturbances of trees, poles, and other tall objects sometimes noticed. Pendulum clocks may stop.

VI   Felt by all, many frightened and run outdoors. Some heavy furniture moved; a few instances of fallen plaster or damaged chimneys. Damage slight.

VII   Everybody runs outdoors. Damage negligible in buildings of good design and construction; slight to moderate in well-built ordinary structures; considerable in poorly built or badly designed structures; some chimneys broken. Noticed by persons driving motor cars.

VIII Damage slight in specially designed structures; considerable in ordinary substantial buildings, with partial collapse; great in poorly built structures. Panel walls thrown out of frame structures. Fall of chimneys, factory stacks, columns, monuments, walls. Heavy furniture overturned. Sand and mud ejected in small amounts. Changes in well water. Persons driving motor cars disturbed.

IX   Damage considerable in specially designed structures; well-designed frame structures thrown out of plumb; damage great in substantial buildings, with partial collapse. Buildings shifted off foundations. Ground cracked conspicuously. Underground pipes broken.

X   Some well-built wooden structures destroyed; most masonry and frame structures destroyed with foundations; ground badly cracked. Rails bent. Landslides considerable from river banks and steep slopes. Shifted sand and mud. Water splashed (slopped over banks).

XI   Few, if any, (masonry) structures remain standing. Bridges destroyed. Broad fissures in ground. Underground pipelines completely out of service. Earth slumps and land slips in soft ground. Rails bent greatly.

XII   Damage total. Practically all works of construction are damaged greatly or destroyed. Waves seen on ground surface. Lines of sight and level are distorted. Objects are thrown upward into the air.

APPENDIX F

A Geologic Overview of Cape Cod: A Field Trip Guide[1]

Introduction

This overview presents a generalized account of the geology of Cape Cod, demonstrated by scenic overlooks from the Cape Cod Canal to the Provincelands dunes. The geology and, in part, the shape of Cape Cod were established by lobation of the Laurentide ice sheet during its retreat. The lobes, from west to east, were the Buzzards Bay lobe, the Cape Cod Bay lobe, and the South Channel lobe, which occupied the western Gulf of Maine east of Cape Cod (Fig. 20). Retreat of the lobes was not synchronous; the Buzzards Bay lobe retreated first and the South Channel lobe, last. As a result, the glacial deposits on Cape Cod, in general, become younger towards the east, and the youngest deposits occur on the outer Cape.

U.S. Geological Survey map I-1736, a geologic map of the Cape and Islands, is a useful aid in understanding the geology seen from the overlooks during the field trip. Ordering information is in Appendix C. Figure 47 in this book is a generalized map of the Cape Cod geology.

Among the earliest visitors to Cape Cod were Gosnold in 1602, Champlain in 1605 and 1606, Hudson in 1609, and John Smith in 1614. There is no evidence that the Vikings were the first to visit, but they could have been. Many of these early visits ended in violence between the Indians and Europeans, including killing and kidnaping. The Pilgrims in 1620 first landed at the tip of Cape Cod. They explored as far south as Eastham, where their first contact with the Indians, a fight, occurred. During the expedition, they stole whatever Indian possessions they wanted. This unhappy situation between the

---

[1] *This field trip guide is modified from U.S. Geological Survey Open-file report 88-329 prepared for the 1988 American Quaternary Association biennial meeting, Amherst, Massachusetts. It contains references to figures in the body of this book.*

Indians and Europeans has a modern counterpart when the natives feel exploited by the summer invaders from offcape.

Cape Cod can be approached in several ways. The Massachusetts Turnpike connects with I-495 south, which ends at U.S. Route 6, the basic route for the trip. U.S. Route 95 connects with U.S. Route I-195 in Providence, Rhode Island, and connects with U.S. Route 6 via State Route 25. From the Boston area, there are two ways to reach Cape Cod. Route 128, the circumferential highway around Boston, connects with Route 24, which connects with I-495. Route 128 also connects with Route 3, which connects with Route 6 at the Cape Cod Canal. Stop 1 is a canal overview from Route 6. The overview is east of Route 25 about 2.0 miles and west of Route 3 about 1.7 miles. From Route 25, it is the third pulloff on the right; from Route 3, it is the second pulloff on the left.

Stop 1. ROUTE 6 CANAL OVERVIEW

The first stop of the field trip is on the east end of the Sandwich moraine and is an overlook of the Cape Cod Canal. The Sandwich moraine and Buzzards Bay moraine form the so-called backbone of Cape Cod. The Buzzards Bay moraine, not visible from this point, parallels the Buzzards Bay shore and forms the Elizabeth Islands chain southwest of Woods Hole. A short distance south of the canal, the Sandwich moraine deposits overlie the Buzzards Bay moraine deposits.

The canal is located in a topographic low that was originally occupied by two streams, the Scussett River that flowed northeast into Cape Cod Bay and the Monument or Manomet River that flowed southwest into Buzzards Bay. The divide between the two streams was at an altitude of about 30 feet. The through valley is thought to have formed as water from a glacial lake in Cape Cod Bay escaped to the Buzzards Bay valley. The outlet may have had an initial altitude of about 80 feet, about the surface altitude of the earliest deltas associated with the lake. Erosion lowered this threshold and consequently the level of the lake. The outlet was abandoned when lower outflow routes developed as the ice retreated from Cape Cod Bay. The through valley and the two rivers provided an easy portage between Cape Cod Bay and Buzzards Bay. By 1627, colonists had established a trading post in the valley for trade between the English settlers of

New England and the Dutch settlers of New Amsterdam. Miles Standish of Alden and Mullens fame proposed a canal between Cape Cod Bay and Buzzards Bay to avoid the dangerous shoals southeast of Cape Cod. More than a century later, a canal was proposed by George Washington, among others, as a way to avoid sea blockades such as those imposed by the British during the Revolutionary War and the War of 1812. After numerous false starts by various companies that were licensed by Massachusetts to dig the canal, the canal was completed in 1914 by a company headed by August Belmont—a feat commemorated by the glacial boulder at this site. Following World War I, the canal was bought by the United States and improved by the U.S. Army Corps of Engineers to what you now see. It is the widest canal in the world.

Leaving stop 1, turn east on Route 6. At the rotary, Route 6 turns right and crosses the Sagamore Bridge. Continue east. Route 6 runs along the crest of the Sandwich moraine, with Cape Cod Bay to the north and Vineyard and Nantucket Sounds to the south. The moraine consists of a veneer of till and boulders underlain by stratified drift, including sand and gravel and rhythmically bedded silt and clay. Both the Sandwich moraine and the Buzzards Bay moraine have been interpreted to be glaciotectonic features formed as the lobes readvanced and thrust previously deposited drift. To the south of the Sandwich moraine lie the inner Cape Cod outwash plains. To the north of Route 6 lie the proximal side of the moraine, deposits of the Cape Cod Bay lake, and marsh and spit deposits of Holocene age.

Continue east on Route 6 to Union Street (exit 8). The distance from stop 1 to Union Street is 22.3 miles. Turn north on Union Street to Route 6A in Yarmouth (distance 1.4 miles). Turn east on 6A. At the cemetery in Dennis (it's on the right), turn right onto Scargo Hill Road (distance 3.5 miles). Shortly after the turn, Scargo Hill Road bears left at the fork with Old Bass River Road. Continue on Scargo Hill Road to a left turn that goes to Scargo Hill tower (Stop 2). Climb the tower for this overview.

Stop 2. SCARGO HILL TOWER

Scargo Hill tower is on one of two kames that rise slightly above the Harwich outwash plain. Glacial deposits located between the kames, the Harwich outwash plain, and the Sandwich moraine and the shore are interpreted to be ice-contact deltas and shallow-water sediments of the Cape Cod Bay glacial lake. The deltas along this part of the shore generally have altitudes of about 60 feet; to the west, older deltas have altitudes closer to 80 feet. The latter may have formed during the earliest high stage of the Cape Cod Bay lake. Initially, the lake was very narrow as the ice pulled away from inner Cape Cod, but it became a significant body of water when the large outwash deltas developed on outer Cape Cod and along the west shore from Plymouth to Duxbury.

Postglacial deposits including spits, beaches, marshes, and dunes formed as the Holocene rise in sea level drowned the glacial cape. Sandy Neck and the Barnstable marshes can be seen to the west. Both features began to develop about 4,000 years ago, when sea level was about 25 feet below the present level. Provincetown Spit, seen at the outer end of Cape Cod if it is a very clear day, started to form about 6,000 years ago.

Return to Scargo Hill Road and turn left. Continue east on Route 6A to Brewster. In Brewster, turn right onto Route 137 (distance 4.6 miles). At Route 28 (distance 7.2 miles), turn left and continue to Chatham. At the rotary (distance 3.3 miles), take Main Street to where it takes a sharp left. Turn right to Chatham Light (Stop 3).

Stop 3. CHATHAM LIGHT

On the way to Chatham Light, the route crosses the much-collapsed Harwich outwash plain. Large and small kettles and shallow valleys interrupt the outwash plain. Many of the large kettle lakes have bottoms well below present sea level and represent ice blocks as much as 200 feet thick. Valleys occur on all of the Cape Cod outwash plains. They are relict because many are dry. They cross kettles and were thus cut early, before the ice blocks buried by outwash melted. These valleys were probably not cut by glacial meltwater because they do not start at the ice-contact head. They may have been cut by runoff of rainwater and water from melting snow before the outwash plains became vegetated. Permafrost could have prevented the surface water from percolating into the sandy outwash plain deposits. However, there is no evidence of permafrost and the outwash deposits themselves provide evidence for abundant meltwater and a temperate climate during deglaciation. An unvegetated or poorly vegetated outwash plain surface may have been sufficient to favor runoff over percolation. Runoff may have been encouraged by the ubiquitous layer of wind-deposited silty sand that caps the coarser, more permeable, outwash sand and gravel. I favor spring sapping for the mechanism that cut the valleys. The valleys are morphologically similar to spring sapping valleys in the West, for example on the Colorado Plateau, and to spring sapping valleys produced experimentally. The valleys may have been cut when the water table was much higher as a result of glacial lakes dammed by the outwash plains.

Cape Cod is a dynamic environment. The dominant factor in the changing landscape is the sea. Where the cape is unprotected from wave attack, sea cliffs have formed in the glacial deposits. In places, barrier beaches protect the glacial deposits from erosion. However, the protection is temporary as the barriers can change form or wash away, especially during major storms. A breach in this barrier at Chatham began during a northeast storm in January 1987 (Fig. 64). The continuing impact of the still-widening breach includes increased tidal range and flow within the lagoon, shoaling of the harbor as flood-tidal deltas develop, and erosion along the mainland shore. Changes in the barrier system protecting the mainland from wave attack appear episodic, as does erosion of the mainland when changes in the barrier system occur. Evidence of a breakdown in the barrier system and erosion of the mainland that occurred in 1851 is shown by the inactive sea cliff below Chatham Light. During the time the Chatham shore was exposed to wave attack, roads and houses were destroyed and as much as 100 feet of cliff retreat occurred. The erosion following the latest breach has already claimed several houses and, unless hard protective measures are taken, will probably claim more. A laissez-faire attitude prevailed for a short time after the recent break, but as expensive land and houses began to go, that changed. The willingness to live with nature gave way to screams for engineering structures to protect the mainland shore from wave attack or to close the breach in the barrier.

To leave Chatham Light, go north on Shore Road and Route 28 to Orleans, where Route 28 joins Route 6A. Continue northward to the Eastham rotary (distance 9.2 miles). Take Route 6 north to Governor Prence Road, a left fork (1.4 miles). Turn right onto Fort Hill Road to Fort Hill overview (Stop 4).

Stop 4. FORT HILL

The glacial features seen from Fort Hill were deposited by the South Channel lobe. From Chatham Light, the route crossed the deeply embayed eastern flank of the Harwich outwash plain. The large embayments, Pleasant Bay and Little Pleasant Bay, represent sublobes of South Channel ice, against which the Cape Cod Bay lobe outwash was deposited.

Fort Hill overlooks Nauset Harbor, an embayment formed by a sublobe of South Channel ice. The Nauset sublobe is thought to have been larger initially and to have occupied the site of the Eastham plain. Fort Hill is part of the Eastham plain; across the harbor, to the southeast, are the Nauset Heights ice-contact deposits. The Nauset Heights deposits may be contemporaneous with, or somewhat younger than, the Harwich outwash plain deposits and are the oldest of the South Channel lobe units. The Eastham plain deposits are considered to be the youngest South Channel lobe outwash unit.

Nauset Harbor is bordered on three sides by sea cliffs cut into the glacial drift. The cliff erosion may have occurred before the barrier spit formed or when the embayment was much larger and the mudflats and marshes had not yet developed. The embayment is the result of the Nauset sublobe and was here when the sea transgressed; the cliffs may originally have been ice-contact slopes that were later modified by wave erosion. The opening in the barrier is presently migrating northward, and the southern spit is about ½ mile longer than it was in 1962. Nauset Harbor spit changed greatly in the 1978 February storm. Before the storm, dunes up to 40 feet high occupied the spit north of the inlet.

Samuel de Champlain visited Nauset Harbor and made the first map of the region in 1605. The map shows the headlands and barrier spits up to a mile seaward of their present position. The distribution of the marsh and sand flats on the map is remarkably similar to the present distribution.

Return to Route 6 and turn right. Continue on the entrance road for the Cape Cod National Seashore headquarters (distance 6.5 miles). Turn right and carefully follow signs for the Marconi site (Stop 5).

Stop 5. MARCONI SITE

The Marconi site is the site from which the first transatlantic wireless transmission between the United States and Europe was made in 1903. Offshore is the site of the pirate ship *Whydah*, wrecked in a storm in 1717 and now being excavated. The ship's bell, cannon, gold, silver, and other artifacts have been found. The presence of the *Whydah* probably explains the gold coins occasionally found in the beach deposits along this shore.

Geologically the Marconi site is located at the southern end of the Wellfleet outwash plain deposits, the largest and next to oldest of the South Channel lobe drift units. Its southern flank was deposited against the Nauset sublobe. Following the retreat of the sublobe, the Eastham outwash plain was deposited against the ice-contact slope of the Wellfleet plain. The difference in elevation between the plains and the general westerly slope of the outwash plain surfaces can be clearly seen. The latter unequivocally establishes the presence of ice to the east and is the primary evidence for ice in the Gulf of Maine during late Wisconsinan time.

The great cliff along the shore from Eastham to Truro demonstrates dramatically the erosion that will eventually reduce the cape to a series of low islands and broad sand shoals. Erosion along this shore, although sporadic and local in nature, averages about 3 feet per year. The eroded material is transported, by longshore drift, southward to nourish the barriers from Eastham to Chatham and northward to build the Provincelands Spit. Major erosion along the east shore of the cape and the southwest and west shore of Cape Cod Bay occurs mostly during severe winter northeast storms. In contrast, major erosion along the south shore and the Buzzards Bay shore occurs during summer and fall hurricanes.

Return to Route 6 and continue north. The road drops down into the Pamet River valley (distance to the Pamet River bridge, 7.6 miles). The Pamet Valley is the largest of the outwash plain valleys (Fig. 39). If not blocked by a barrier beach on the Atlantic end, the Pamet River would be a seaway, completely separating the lower cape into two parts and making the northern part an island. Its size and length indicates that it may have had a somewhat different origin than the other outwash plain valleys. Possibly, the Pamet Valley may have been enlarged when, through headward erosion, it completely breached the Wellfleet plain to drain a lake that existed to the east of the plain. The other valleys in the Wellfleet outwash plain, called pamets or hollows, are probably the large end member of all the outwash plain valleys. Within the Wellfleet plain, they represent significant fluvial erosion shortly after the plain was built. Where the hollows interrupt the great cliff, they provide access to the beach. They were a vital means of escape for sailors stranded on the beach and also made it easier for wreckers to save lives and then scavenge the wrecks.

Continue north on Route 6 and turn right onto South Highland Road (distance 2.5 miles). Go to Highland Light Road and Highland Light (Stop 6).

Stop 6. HIGHLAND LIGHT

One of the early classic Quaternary sections in New England is exposed in the cliffs below Highland Light. The section consists of rhythmically laminated clay that is underlain by gravel and overlain by fine sand. The deposits are glaciolacustrine in origin, laid down in a glacial lake, unrelated to the Cape Cod Bay lake, that developed between the Cape Cod Bay and South Channel lobes and the ice-contact flank of the Wellfleet plain.

Three late Wisconsinan outwash plains can be seen from this over-look. You are standing on the surface of the lake deposits that form the Highland plain (Fig. 47). The ice-contact slope of the higher Wellfleet plain can be seen to the south, marked by the granite tower, which is a memorial to Jenny Lind, the famous Gay-Nineties song bird. To the north is the lower Truro plain. The Wellfleet and Truro plains were formed as deltas, deposited into different levels of the Cape Cod Bay glacial lake. The Wellfleet plain is the oldest, and the Truro plain is the second youngest feature related to the lake. The Eastham plain, to the south of the Wellfleet plain, is considered to be the youngest because, above sea level, its deposits are entirely fluvial; and the plain was graded to a lower and, most likely, younger level of the lake.

Highland Light is the most powerful light on the New England coast and can be seen from more than 20 miles at sea. The light is threatened by the retreat of the cliff, which has averaged about 5 feet per year during the last 10 years. The retreat is facilitated by landslides in the lake clay and overlying sand.

Return to Route 6 by taking a right onto South Highland Road and a left onto Highland Road. Turn north on Route 6. At High Head (1.9 miles), the road crosses from glacial deposits to the deposits of the Provincelands Spit. Continue northward to Point Road (distance 4.5 miles). Turn right into the Provincelands Visitor Center (Stop 7).

Stop 7. PROVINCELANDS VISITOR CENTER

The route from Highland Light to the final overview leaves the glacial part of Cape Cod at High Head. The scarp just east of Pilgrim Lake is a sea cliff that was eroded in the glacial drift before the Provincelands Spit was formed. Just beyond Pilgrim Lake, a lagoon now artificially closed from the sea, the dune sand is burying trees (Fig. 70) and encroaching on the road. The sand must be removed to keep the road open. Dark layers within the dune sand are old soil horizons (Fig. 59). The Provincelands spit began to form about 6,000 years ago as the sea transgressed the glacial cape. Longshore currents carried sand eroded from the glacial deposits northward and westward into deep water off the tip of glacial Cape Cod to form a recurved spit. As sea level rose, new spits were developed along the Atlantic shore, each one recurving toward Cape Cod Bay. The oldest beach deposits occur along the Cape Cod Bay shore and they become progressively younger toward the Atlantic Ocean. As the barrier beach grew toward the Atlantic Ocean, erosion along the Cape Cod Bay side has removed the oldest spits, at least in part. Wind erosion and deposition continue to move and modify the dunes. Except for the foredune along the Atlantic shore, the dunes may be unrelated in age to the beach deposits beneath them.

Although Cape Cod is a fragile land, its ability to adjust to sea-level rise is well shown by the developing spits and shoals. If sea level continues to rise, Cape Cod may persist in some form long after the rock-bound coast of New England and its port cities are drowned.

To leave Cape Cod, return to Route 6 and turn left.

APPENDIX G

Other Books on the Geology and Natural History of Cape Cod and the Islands

Allen, E. S., 1976, A wind to shake the world: Little, Brown and Company, Boston, 370 p.

An exciting account of the 1938 hurricane and its impact on New England. Leaves a lasting impression of the danger of the storm surge. If you have property in the flood plain, you should read it.

Chamberlain, B. B., 1964, These fragile outposts: Doubleday. Reprinted, Parnassus Imprints (1981), Yarmouth Port, Massachusetts, 327 p.

A thorough description of the geology and parts of the natural history of the Cape and Islands. Some of the geology is outdated.

Emery, K. O., 1969, A coastal pond: American Elsevier Publishing, New York, 80 p.

The natural history of an upper Cape coastal pond studied by oceanographic methods.

Hale, Anne, 1988, Moraine to marsh: Watership Gardens, Vineyard Haven, Massachusetts, 195 p.

A well-illustrated and easily read field guide to the natural history of Martha's Vineyard.

Imbrie, John, and Imbrie, K. P., 1979, Ice ages: Solving the mystery: Enslow Publishers, Hillside, New Jersey, 224 p.

A fascinating and readable account on understanding the causes of the growth and decay of global ice sheets.

Leatherman, S. P., 1988, Cape Cod field trips: From yesterday's glaciers to today's beaches: The University of Maryland, College Park, Maryland, 132 p.

A well-illustrated field guide to the geology of Cape Cod. Stresses the development of the shoreline.

O'Brien, Greg, ed., 1990, A guide to nature on Cape Cod and the Islands: Viking Penguin, New York, 240 p.

Chapters written by local authorities on the natural history of the Cape and Islands. Answers many commonly asked questions and is easy to read.

Oldale, R. N., 1981, Geologic history of Cape Cod, Massachusetts: U.S. Geological Survey Popular Publication, 23 p.

A color booklet about the geological history of the Cape. Simply written.

Raymo, Chet, and Raymo, M. E., 1988, Written in stone: Globe Pequot Press, Chester, Connecticut, 163 p.

An easily understood and up-to-date account of the geological development of the northeastern United States from Precambrian times to today.

Redfield, A. C., 1980, Introduction to tides: Marine Science International, Woods Hole, Massachusetts, 108 p.

A somewhat complex, but understandable discussion of the tides of New England and New York. It includes detail on the tides around Cape Cod and the Islands.

Strahler, A. N., 1966, A geologist's view of Cape Cod: Doubleday. Reprinted Parnassus Imprints (1988), Orleans, Massachusetts, 115 p.

The classic account on the geology of Cape Cod. Simple and easy to read. It addresses mostly the physical geology of the Cape.